Five Technologies for Educational Change

- Systems Thinking
- Systems Design
- Quality Science
- Change Management
- Instructional Technology

David F. Salisbury

Educational Technology Publications
Englewood Cliffs, New Jersey 07632

Library of Congress Cataloging-in-Publication Data

Salisbury, David F.
   Five technologies for educational change : systems thinking, systems design, quality science, change management, instructional technology / David F. Salisbury.
      p.   cm.
   Includes bibliographical references and index.
   ISBN 0-87778-293-8
   1. Educational technology--Philosophy.   2. Educational innovations.   3. Educational change.   4. System analysis.   5. Instructional systems--Design.   6. Computer managed instruction.   7. Computer-assisted instruction.   8. Educational evaluation.   9. Education--Computer network resources.   10. Total quality management.   I. Title.
LB1028.3.S264   1996
371.2'0285--dc20   96-3650
                                                                                          CIP

Copyright © 1996 by Educational Technology Publications, Inc., Englewood Cliffs, New Jersey 07632. All rights reserved. No part of this book may be reproduced or transmitted, in any form or by any means, electronic or mechanical, including photocopying, recording, or by any information storage and retrieval system, without permission in writing from the Publisher.

Printed in the United States of America.

Library of Congress Catalog Card Number:
96-3650.

International Standard Book Number:
0-87778-293-8.

First Printing: May, 1996.

ACKNOWLEDGMENTS

For the past 12 years I have had the opportunity to work on a number of educational change projects. These projects, and the people associated with them, were the source of much of the material and ideas that became embodied in this book. I am grateful to all of the people who have shared with me their knowledge and expertise on so many subjects, and for their intellectual stimulation and support. I am particularly indebted to Bob Branson, Director of the Center for Educational Technology; Wendy Cullar, Program Director for The Florida Schoolyear 2000 Initiative; Bob Morgan, Director of the Learning Systems Institute, and Marty Beech, Pam Zimpher, Donna Gillis, Owen Gaede, and Craig Johnson, all at the Center for Educational Technology.

The Florida Schoolyear 2000 Initiative, the large-scale school reform project on which I worked at the Center for Educational Technology, provided me with the resources, knowledge, experiences, and personal contacts that enabled me to produce this book. For this, I must thank all of the project's participants, sponsors, and development partners, including the Florida Department of Education, the Institute for Simulation and Training at the University of Central Florida, Information Systems of Florida, ODR Inc., the Training Systems Division of the Naval Air Warfare Center, Technical Assistance Research Programs, American Institutes for Research, Encyclopaedia Britannica Educational Corporation, American Management Systems, Chancery Software, and BBN Corporation.

One of the many valuable experiences that The Florida Schoolyear 2000 Initiative provided was the opportunity to attend a three day executive training session in Atlanta with ODR Inc. This intensive workshop gave me my first exposure to the science of "change management." I am particularly grateful to Daryl Conner, President and CEO of ODR Inc., for allowing me to draw extensively (in Chapter 8) from an article we coauthored for *Educational Technology* magazine. That project provided a timely opportunity to describe many of the ideas that are summarized in this volume.

My ideas have also been influenced by people such as Jack Bowsher, former director of external education programs at IBM; Lew Perelman, formerly of the Hudson Institute and now at Discovery Institute; Chester Finn, former Assistant Secretary of Education and Professor of Education and Public Policy at Vanderbilt University; Lou Kompare, former Director of Technology for Walt Disney World; Captain Ernie Lewis, former commander of the Naval Training Systems Center in Orlando, Florida; Jay Forrester, Professor Emeritus of Systems Dynamics at MIT, and Peter Senge, Director of the Systems Thinking and Organizational Learning Program at MIT's Sloan School of Management.

I have also benefited greatly from my conversations with David Pesanelli about planning and design; with Gloria Gery about electronic performance support systems; with Frank Caplan about quality systems, and with David Bayless about quality teams in schools.

My research was aided substantially by a series of seminars on systems design in education that I attended at Pacific Grove, California, sponsored by the International Systems Institute. At these conferences, Bela Banathy, a great friend and mentor, illuminated many of the core concepts discussed in Chapters Two and Four on systems thinking and the design of educational systems. I am also indebted to the members of the Systemic Change Division of the Association for Educational Communications and Technology for their individual writings and contributions to the ideas behind systems design in education, particularly Charlie Reigeluth, at Indiana University and President of the Systemic Change Division, and Kyle Peck, at The Pennsylvania State University.

I also gained a great deal from a conference on redesigning schools that I attended in 1987 in Snowmass, Colorado, where I first met West Churchman, author of several of the core works on systems thinking, such as *The Systems Approach* (1979, Dell) and *The Systems Approach and Its Enemies* (1979, Basic Books). Bela Banathy's and Albert Shanker's speeches at that conference were a starting point for many of the ideas that eventually became part of this book.

I also wish to express appreciation to Stan Marshall of the James Madison Institute for Public Policy Studies in Tallahassee, Florida for allowing me to spend a semester there as a Visiting Scholar. The experience allowed time for completing this book, and I was enriched by the intellectual environment of the Institute and the ideas of the many people associated with it.

I especially appreciate the generous, detailed, and creative comments I received from colleagues and friends who critiqued and made suggestions

Acknowledgments

on several of the chapters. These include Bob Richards, Idaho National Engineering Laboratory; Dale Lick, Learning Systems Institute at Florida State University; Donna Gillis, Fanchon Funk, and Craig Johnson, all at the Center for Educational Technology; Scott Broetzmann, Technical Assistance Research Programs, Inc.; Gene Hall, University of Northern Colorado at Greeley; Sandy McCarroll, The Edison Project; Jerry Kemp, San Jose State University; Rodney Earle, Brigham Young University; and Kurt Rowley, Florida State University. I am deeply grateful to Loralee Wooten for her help with editing and proofreading. She made many valuable comments and suggestions that greatly improved the book. I also want to thank the graduate assistants who helped in the fact finding associated with writing this book: These are LeGrand Hayes, Eric Jakubowski, Ron Ricks, and Craig Loveland. Thanks go to Larry Darnell, Jean VanDyke, and Kay Cranford, who assisted with production of the graphics.

A most sincere appreciation goes to Lawrence Lipsitz, President of Educational Technology Publications, not only for his guidance and encouragement, but also for his input into the content and structure of the book. His insights and suggestions have made the book much more valuable than it would have been otherwise.

Lastly, I must thank my wife Elaine for reading through each chapter with me and helping to shape ideas, strengthen certain points, tone down where necessary, or personalize things more. Thanks, Elaine, for your help, encouragement, and perseverance. Thanks for sharing my enthusiasm for the ideas expressed in the book.

Although the book has been greatly improved by the help and contributions of the individuals and groups I have mentioned, my acknowledgment of these people and organizations should not, of course, be taken to imply any endorsement. The opinions and views expressed here are entirely my own.

PREFACE

This book is not about education and schooling the way it is today. Rather it is about the way education could be and the Five Technologies that are essential to making it that way. By "technologies," I don't mean just machines or electronic tools—I mean the application of theory and technique that must be studied and mastered to be put into practice. The Five Technologies represent five essential knowledges or disciplines that can and should have a profound impact on education. Each of these Five Technologies has been revolutionary in its own right. Each is used extensively in the most successful businesses and industries, but unfortunately, little in schools. One or two of them may have found their way into a school or school district here or there, but generally they have been applied as isolated strategies. They have not been used in the most productive way because they have not been viewed as mutually dependent. They cannot be used successfully, I suggest, unless they are inseparably fused with the others.

Over the last two decades, a virtual revolution has taken place in America's corporations and businesses as they have had to compete with new global competitors, find new ways to optimize and economize, and improve quality. They have had to reinvent themselves, restructure their organizations, reengineer their processes, and adopt and improve on techniques and technologies borrowed from their overseas competitors. The same type of revolution must take place in our schools if they are to effectively serve the needs of the future. The Five Technologies provide the keys to that revolution.

Increased competition, choice, and consumer demands are affecting education today in the same way that these factors have affected American business and industry during the past two decades. Schools will no longer be able to rely on regulation and standardization. Instead, they will need to learn to thrive in an environment of competition and continuous improvement. Creativity, innovation, modernization, and customer service will become the keys to success in the future.

One of the greatest weaknesses of school practice is the "not invented here" attitude, our disinterest in new ideas from beyond the school walls—ideas from outside of education. We seem to ignore the fact that the education enterprise has much in common with other service industries such as hospitals, corporate training, information services, and telecommunications. The revolution in education will require the accumulation of a broad range of technologies and techniques invented and developed by others. Our modern society will require that organizations (including schools) search for new ideas and capitalize on them quickly. The Five Technologies represent some of the most important new ideas to surface during the past two decades. They offer powerful new tools for improving education.

The Five Technologies, when viewed together as a unified strategy for reform, represent a major expenditure of effort. But only when these Five Technologies converge will we be able to create and deliver the kind of educational experiences students and parents want and will need for the future.

NEW ORGANIZATIONS FOR LEARNING

I had a hard time choosing the title for this book. Initially, I used the word "school" in the title because I felt that the audience would resonate best with that term. But the book is really about more than just restructuring or reforming schools—it is about how to create completely new kinds of organizations to foster learning. Occasionally in the book I use the term "school restructuring." When I use this term, I do so only because it may be the first step toward more efficient and effective means of providing education services to students and families in the future. The ultimate goal, however, is not just to reinvent schools, but to invent entirely new ways to meet people's learning and education needs.

As technology increases, learning in school facilities may take on less and less importance. Electronic learning resources, telecommunications, computer simulations, and the entire technology of the emerging information infrastructure will make learning better, faster, and cheaper and will allow it to take place easily at home or in a variety of other individual or group settings. New organizations for learning in the future may bear little resemblance to traditional schooling.

School restructuring today must reflect what learning might be like tomorrow. If we are to redesign schools, we should design them so that they become fluid organizations that will change and adapt easily to new cir-

Preface

cumstances. If restructured schools are to be a reflection of the learning methods of tomorrow, they will at least need to:

- become more effective and efficient by eliminating methods and practices that don't work well;
- become more customer-oriented by giving learners and parents a variety of options and choices (this means not only allowing parents to decide which schools their children should attend, but also to have some choice of individual teachers, curriculum, and learning methods);
- provide students, parents, and members of the community access to electronic knowledge-bases and information services;
- allow students to change roles from that of passive recipients of information to active information workers (this means students can access appropriate information and use it to write reports, draw conclusions, or solve problems);
- provide instant, continuous, electronic communication between students, teachers, parents, and home; and
- replace the concept of "classroom" (with its rows of seats and desks) with something that resembles a modern automated office environment (with individual and group work areas and high tech learning and communication resources).

MY BACKGROUND

My career has provided me with marvelous opportunities to observe education from a variety of perspectives. For more than a decade, I had the opportunity to work on educational change projects at the Center for Educational Technology. What I have learned there is due to the Center's long-term focus on training, human performance, and education in large organizations. In my work, I had the privilege of being involved in and observing a number of major projects in industry, the military, government, and education in the U.S. and in other countries. In all of these projects, the goal has been to use technology and scientific principles to improve human performance. The experiences have allowed me to learn about the Five Technologies, provided me with excellent sources and access to information, and afforded me invaluable interaction with colleagues and experts in the various disciplines. This intellectually rich yet pragmatically oriented setting has produced the knowledge needed to write this book.

I did not, of course, invent the Five Technologies I am writing about in this book. The Five Technologies described here are based on the results of research, experience, writing, theorizing, and invention by hundreds of individuals. Their work has produced today's state-of-the-art developments in these technologies. I have been fortunate enough to be in a position to see these Five Technologies put into practice over a period of several years. I have been able to refine my ideas about them, discuss them with colleagues, collaborate on research, and introduce them to schools and to educators. What I have tried to do in this book is to combine and present these ideas in a way that is understandable and practical for others.

WHY I WROTE THIS BOOK

My wife and I are the proud parents of six children. Currently, we have one child in high school, one in middle school, one in elementary school, and a preschooler. We also have a middle school age child and an elementary age child that we teach at home. We chose to begin teaching our children at home during certain years because we could see that the local public schools were not able to provide the kind of education we wanted for our kids. The schools seem to have lost the ability to teach well the basic skills of reading, writing, mathematics, history, geography, and science. Instead, they have become bogged down in bureaucratic jargon, platitudes, and subjective outcomes. We wanted our children to acquire the basic and advanced skills they will need for college and for later life. Teaching them at home, at least for a part of their schooling career, seemed to be their best chance. Admittedly, not all parents can or want to do this, and that is part of my motivation for writing this book. Because I care about children and families, I want to suggest how schooling might be re-created to enable more children to get a better learning experience and how parents can take a larger role in the education of their children. I still believe that education is the key to a productive life. I believe that the Five Technologies described in this book are the keys to creating an educational enterprise that can help students obtain what they will need for the future.

WHAT YOU WILL GAIN FROM THIS BOOK

I believe that when you write a book, it is a good idea to tell your readers what they will get out of reading it. What you will get out of this book will depend on the role you now play in the current education enterprise. If you are an education leader responsible for change in a public or private school or school system, this book can guide your efforts in directions that will make them truly productive. The knowledge embodied in the Five Technologies has created successful, customer oriented, competitive businesses and industries all over the world. This knowledge is essential if you are to obtain the valuable outcomes you seek for your schools.

Preface

If you are a policy maker, legislator, school board member, or a reform advocate, this book will provide a readable summary of the Five Technologies. The Five Technologies have had a major impact on business and our economy. However, knowledge about these Five Technologies is currently available only from high priced consulting groups or specialized business management books and often is not explained well to a general audience. Thus, this book can help you "catch-up" with these important new developments.

If you are a teacher or administrator who is currently involved in school, district, or state-wide restructuring, you will find the book useful as you experience the process. It will give you a new set of "lenses" through which you can view the experience, and perhaps help you identify some gaps or weaknesses in what you have been or are now doing.

If you are a consultant, management specialist, educational technologist, or other professional working to improve our current approach to education, this book will help you apply the tools and technologies that are required for improving the total system. The Five Technologies are a common sense paradigm that can help assure that your efforts will have an overall impact and lasting positive effect on the system as a whole.

If you are a business executive who is concerned about the state of education in this country, this book will provide you with essential knowledge about how the current educational system can and must change. More importantly, it will help you pinpoint strategies for becoming involved in effective business-school partnerships and show how your business's collective expertise could be channeled toward helping education. The quality sciences and the modern management methods that business leaders have incorporated into their daily work must be migrated to schools. This is the critical role you can play.

If you are a parent, taxpayer, or just a thoughtful citizen, you will gain information about what kind of educational system we want for the future and how we can begin to create that future. Those who are outside the education establishment have an important and legitimate role to play in helping to improve education. In the same way that war is too important to be left solely to the generals, governing solely to the politicians, business solely to the managers, or hospitals to the physicians, education is much too important to be left solely to the school professionals. The generals, politicians, managers, physicians, and educators are constantly tempted to run the enterprise for themselves instead of for the students and parents, the voters, the taxpayers, the patients, or the other people they serve. Not intentionally, of course, but only because humans tend to react naturally in that way. It is difficult for those working inside an organization to see things from a larger perspective.

It is for this reason that change in education will most likely not come from within but from without. There is a traditional culture in schools that influences the behaviors and assumptions of all those who work there. Those who are working within the education enterprise are naturally embedded in the self-interest of this culture. Outside reformers, innovators, and visionaries will recognize that this culture exists. They will wisely look for opportunities that challenge that culture; but to be successful, they must not totally threaten it. The Five Technologies presented in this book challenge the current culture in schools, but they do not ignore its existence. If they did, they would be destined to be shunned and dismissed as irrelevant.

Finally, this thought—the success or failure of our schools affects all of us. To adapt what Leon Trotsky said of war: You may not be interested in schools, but schools are interested in you. Our schools provide our society with many benefits. If the schools work well, our children receive the skills and knowledge they need to succeed in their careers and contribute to our economy, health, and happiness as a nation. Schools should also provide our children with the dialogue of democracy, of free enterprise, of rights and responsibilities, and of character traits that enable them to succeed as adults. America's prosperity, economy, and general welfare, therefore, depend on the success of our schools. School leaders, citizens, politicians, and others who are currently working to reform our education system should remember that all of us will be affected by the outcome.

CONTENTS

Acknowledgments v

Preface ix

Chapter 1: Introduction to the Five Technologies 3

PART I
THE FIRST TECHNOLOGY:
SYSTEMS THINKING

Chapter 2: Systems Thinking 17
Chapter 3: Systems Thinking Tools 33

PART II
THE SECOND TECHNOLOGY:
SYSTEMS DESIGN

Chapter 4: Systems Design 45
Chapter 5: Systems Design Tools 77

PART III
THE THIRD TECHNOLOGY:
QUALITY SCIENCE

Chapter 6: Quality Science 93
Chapter 7: Quality Science Tools 127

PART IV
THE FOURTH TECHNOLOGY: CHANGE MANAGEMENT

Chapter 8: Change Management 147
Chapter 9: Change Management Tools 169

PART V
THE FIFTH TECHNOLOGY: INSTRUCTIONAL TECHNOLOGY

Chapter 10: Instructional Technology 183
Chapter 11: Instructional Technology Tools 215

Chapter 12: The Five Technologies Revisited 229

Index 235

FIVE TECHNOLOGIES FOR EDUCATIONAL CHANGE

Systems Thinking
Systems Design
Quality Science
Change Management
Instructional Technology

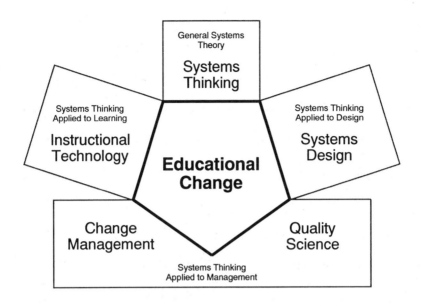

CHAPTER 1

INTRODUCTION TO THE FIVE TECHNOLOGIES

Technology is not just something you plug into a wall outlet.

Educators today find themselves on the verge of a revolution—a revolution of methods in the public schools. Ever since the publication of *A Nation at Risk* (the 1983 report by a special White House Commission) educators and others interested in education have proposed numerous recommendations for reform. In 1992, President Bush initiated the New American Schools Development Corporation, which funded a competition for design teams to invent "a new generation of American schools"—and called for schools to break out of the traditional mold. The Clinton Administration adopted Goals 2000, and is pushing for national education standards and other reforms. Whittle Communications' "Edison Project" has set out to establish a chain of schools that incorporate innovative curricula and technology. School districts and states throughout the country are initiating efforts to restructure their education systems. Politicians and public advocacy groups are proposing voucher, school choice, and charter school programs that will likely have a profound impact on public education and increase the need for public schools to compete in terms of quality and productivity.

In many ways, what is happening in education today is similar to what has already occurred in corporate America as our businesses and industries have faced increased competition in a global marketplace. Even the best American companies have been forced to:

- rethink the way their organizations work and the environment in which they operate;
- learn about "systems thinking" and give up the illusion that parts operate as separate, unrelated forces;
- design whole new organizational structures;
- reengineer their processes;
- adopt quality systems and "total quality management" in order to meet increased customer demands;
- manage and painfully absorb a great amount of change during a short period of time; and
- use modern electronic tools to increase productivity and meet customer demands.

We are just now beginning to wake up to the fact that our schools will need to experience this same painful process and reinvent themselves, if they are to continue to serve the needs of America in terms of quality, inventiveness, and results. Schools must look at what businesses have gone through and how they have accomplished it. These observations will give us a big clue about what schools will need to do and how long it may take.

Simply making small changes for change's sake will not be enough. Already some schools and districts are in dangerous straits for having restructured, reorganized, and blindly installed site-based management programs, computer-labs, or a thematic curriculum in the belief that these were some sort of "magic bullets" for improving results. These small reforms are akin to oiling the parts of a worn-out piece of machinery. What is needed instead is to transform our current education system into a totally different kind of enterprise. Such a transformation will require creative and productive thinking, planning, and work. I am writing this book because I believe that the major changes necessary for a meaningful redesign of education can only take place through the use of five new "technologies." In these Five Technologies lies the power to effectively reinvent our educational systems and to create systems that will better meet the needs of the modern world.

THE FIVE TECHNOLOGIES

All of these Five Technologies are used extensively in business and industry but very little in schools. Each of these technologies has been developed separately over the last few decades on a frontier that crosses national

Introduction to the Five Technologies

boundaries. These Five Technologies have enabled industry leaders around the world to effectively remold their organizations into customer oriented businesses that can compete effectively in the global environment. Together the Five Technologies provide the tools for effective restructuring of our educational enterprise.

The First Technology: *Systems Thinking*

Systems thinking allows us to avoid being continuously distracted by each passing educational fad. It allows us to see which changes or improvements will have a powerful overall effect and which will likely be impotent or even counter-productive. Without systems thinking, we can often make worse what we intend to make better. To make real improvements, we have to see education through a correctly focused "systems lens."

The Second Technology: *Systems Design*

Systems design is the technology that enables us to make change and improvement a conscious process. Most changes in education have occurred in tiny amounts and have affected only a small portion of the total student population and education environment. Most of these changes have been piecemeal. That is, they changed only one aspect of the education process at a time, making a small improvement here, another there. In today's world of rapid change and increasing expectations, such a piecemeal improvement strategy is not enough. When a system has not or cannot keep up with changing demands, it has to recreate itself—or die. Systems design gives us the tools to create a whole new system and to plan a coherent strategy for change.

The Third Technology: *Quality Science*

Quality science is the technology of producing a product or service that meets the customer's demands and expectations. Quality science is comprised of a set of scientific and mathematic principles that allow us to continuously improve our processes, obtain a good return on the investment of our human or other resources, and to "do-it-right-every-time." Quality science has been the most valuable tool in the revolutionary change that has occurred in business and industry, but it is used very little in education. It is one of the Five Technologies that are crucial for meaningful school restructuring.

The Fourth Technology: *Change Management*

When people try to implement one or two of the technologies without the others, a great deal of frustration results and negative energy is expended. Negative energy is created whenever people are trying to work against each other (where 1 + 1 equals ½). It causes people to criticize others, to engage in politicking, to bias information, and to protect their own turf. People have hidden but skillful ways of sustaining old behaviors that thwart efforts to learn new ones. Change management is a way to guide creative energy toward positive change.

The Fifth Technology: *Instructional Technology*

Instructional technology actually consists of two parts—electronic learning tools (computers, computer networks, multimedia, telecommunications) and the instructional *design* methods and strategies that are necessary to make these electronic tools effective. These electronic learning tools are changing and will continue to change the way we communicate and learn. We will one day come to look at the idea of a student working only with a pencil, textbook, and worksheet as passé, obsolete, and counterproductive. There will be better, quicker, easier, and cheaper ways to acquire knowledge and skills.

Electronic technologies have already remolded most businesses and human service activities into more productive, customer service oriented enterprises, and they are starting to become more critical to schools. But electronic technologies have only improved businesses or schools when those organizations have first gone through a basic systems re-engineering of the processes to be automated. Automating the old processes produces little, if any, positive effect. This is why instructional technology must be viewed as part of a larger strategy that includes the other four technologies as well.

All of the Five Technologies are related to some degree. The First Technology (systems thinking) represents the general concept of systems—the idea that things are related and interconnected. As you will see, *systems thinking is the glue that holds everything else together.* The other four technologies are derived from systems thinking:

- *Systems design* is the technology of designing and building completely new systems.
- *Quality science* is the application of systems thinking to management and to the problem of producing a product or service that meets customer satisfaction each time, every time.
- *Change management* is systems thinking applied to the management aspects of orchestrating innovation and change.
- *Instructional technology* is systems thinking applied to instruction and learning.

Together, these Five Technologies represent the most recent thinking in the areas of organizational design, learning, instruction, and management.

DEFINITION OF "TECHNOLOGY"

It may seem odd to refer to the five areas listed above as "technologies." Most people are probably used to thinking of "technology" only as machines or electronic devices that you plug into the wall. But the formal definition of technology is "the systematic application of scientific or other organized knowledge to practical tasks."[1] In other words, when we develop a new product or innovation, the disciplines, practices, procedures, tools, and techniques that converge to make up the new innovation are called technologies. Although we tend to think of technology as the machines that produce a new innovation or product, the machines themselves are no more important than the disciplines, practices, procedures, and techniques that go into planning and designing the product in the first place. The combination of disciplines involved in producing the product are just as much technologies as are the machines that are used. (See the insert below.)

The Five Technologies, then, are five essential disciplines that must converge to solve the problems of education. Together, the Five Technologies deal with the combination of tools, machines, electronic devices, people, processes, and management skills that are needed for effective school restructuring. The Five Technologies emphasize that the power for effective educational change is not in new hardware or electronic tools (although these are a necessary part). Rather, it is the *combination* of the intellectual content, management skills, processes, and new tools inherent in these Five Technologies that provide the power for effective change.

DEFINITIONS OF 'TECHNOLOGY'

Technology is often confused in the public mind with electronic devices and complicated machines. The word technology, as used by scientists and philosophers of science, refers to the means by which we make science work to solve practical problems. There may or may not be machines involved in technology, but there is always the application of science. Technology, in its truest form, refers to endeavors to solve human problems. It is a way to manage people, events, and machines by using known and proven tools, procedures, and techniques. Here are some examples of statements that reflect this use of the word technology:

"Technology is a rational discipline designed to assure the mastery of man over physical nature, through the application of scientifically determined laws."

— Y. R. Simon, Pursuit of Happiness and Lust for Power in Technological Society. In C. Mitcham & R. Mackey (Eds.), **Philosophy and Technology** (New York: Free Press, 1983).

"Technology, in its concrete, empirical meaning, refers fundamentally to systems of rationalized control over large groups of men, events, and machines by small groups of technically skilled men operating through an organized hierarchy."
> —J. McDermott, Technology: The Opiate of the Intellectuals. In A. H. Teich (Ed.), **Technology and Man's Future** (New York: St. Martin's Press, 1981).

"The word technology (in the Latin form it is 'texere,' to weave or construct) does not necessarily imply the use of machines, as many seem to think, but refers to 'any practical art using scientific knowledge.' This practical art is termed by the French sociologist Jacques Ellul as 'technique.' He believes that 'it is the machine which is now entirely dependent upon technique.'"
> — Paul Saettler, **A History of Instructional Technology** (New York: McGraw-Hill, 1968).

"In addition to machinery, technology includes processes, systems, management, and control mechanism both human and non-human, and . . . a way of looking at the problems as to their interest and difficulty, the feasibility of technical solutions, and the economic values—broadly considered—of those solutions."
> —James Finn, Technology and the Instructional Process, **AudioVisual Communication Review**, Vol. 8, No. 1, 1960.

"Science dwells on discovering true facts, and relationships of observable phenomena in nature, and with established theories that serve to organize masses of verified data concerning those facts and relationships. In contrast, technology deals with tools, techniques, procedures: the artifacts and processes fashioned by modern industrial man to increase his power of mind and body . . . Technology is action rather than the pure thought that is science."
> — Admiral Hyman Rickover as quoted in S. J. Knezevich and G. G. Eye (Eds.), **Instructional Technology and the School Administrator** (Washington, DC: American Association of School Administrators, 1970).

"Technology refers to endeavors to solve human problems. The central distinguishing characteristic between science and technology is a difference in goal—science to understand the natural world and technology to make modifications to meet human needs."
> — Excerpted from the **Science Education Standards**, National Science Foundation, November 1994 draft.

"**Technology** \tek'näl_ji\ *noun* [Greek *technologia*, systematic treatment] **1** the science of the application of knowledge to practical purposes."
> — **Webster's Third New International Dictionary of the English Language, Unabridged,** 1981.

AN ENSEMBLE OF TECHNOLOGIES

The Five Technologies must be viewed as an ensemble rather than as separate entities. Each of the technologies is only truly effective when it is used in conjunction with the others. In the same way that the instruments in a musical ensemble are selected to form the desired musical result, the Five Technologies are totally complementary and interdependent. The individual musical instruments playing alone, without the accompaniment of the others, produce nothing like the total effect. In the same way, the Five Technologies, if applied separately, produce only a small, fragmented effect. The power for positive change comes through their complementary forces and actions.

Peter Senge, in his book *The Fifth Discipline*, uses the example of the DC-3 airplane to show how diverse technologies are required for producing a major innovation. Production of an airplane requires a combination of physics, metallurgy, chemistry, electricity, manufacturing processes, management techniques, technical training, quality control, and other technologies. These technologies may come from isolated developments in diverse disciplines, but at the point of innovation they become "an ensemble of technologies that are critical to each others' success." "Until this ensemble forms," Senge points out, "the idea, though possible in the laboratory, does not achieve its potential in practice."[2]

Applying the Five Technologies all at the same time is difficult; it is much easier to apply new tools separately. Integrating them as a whole requires discipline and careful orchestration. But it is the combining of the tools that produces the high payoffs. The Five Technologies gain their greatest power only when they are working together. If you use only one or two of these technologies at a time, to employ yet another metaphor, it is like playing a card game without a full deck. Without all Five Technologies, we do not have the essential pieces that must fit together to bring about the overall result.

SYNERGISTIC RELATIONSHIP OF THE FIVE TECHNOLOGIES

The word "synergistic" describes the dynamic, interdependent relationship that exists between the Five Technologies. "Synergy" refers to relationships that build on each other. If we consult the dictionary, we find the following definition for the word *synergy*: "Cooperative action of discrete parts such that the total effect is greater than the sum of the two or more effects taken independently." Simply stated, you have synergy when two or more things produce an effect that they could not produce unless they were acting together.

In a synergistic relationship, 1 + 1 is greater than 2. In the synergistic union of the Five Technologies, 1 + 1 + 1 + 1 + 1 is greater than 5. Working in harmony, they produce an effect much more powerful than they could possibly produce alone. Synergistic is the opposite of antagonistic, where individual parts work in hostility or opposition to each other.

Some people have tried to implement one of the individual technologies in education without using the other four technologies. This is virtually hopeless. Each of the Five Technologies needs the effect of the others to work well. Systems thinking provides the ability to visualize how things work together to produce the overall effect. Systems design provides the process for creating new organizations, programs, and processes. Quality science focuses the system on the customer and on consistent results that meet predetermined quality levels. Change management provides the skills to implement and sustain large-scale change. Instructional technology applies what we know about learning, instruction, assessment, and instructional media. To reach its individual potential, each technology needs the forces provided by the other four technologies. Unless all five of the technologies are present and acting together harmoniously, the individual technologies become separate gimmicks or individual fads. The critical component in school restructuring is to look at how these five powerful technologies interrelate and work together. The task is to integrate them as a whole through careful management and orchestration.

The Five Technologies are like the juggler's ninepins—you have to keep them all in the air at the same time. Juggling requires a great deal of manual dexterity. Juggling the Five Technologies requires an equal amount of *management dexterity*—a constant manipulation of these Five Technologies to achieve a desired end. This is the challenge for education leaders and reformers of the next decade.

In any organization, there exist individual, group, and organizational counterpressures that will limit the success of any change effort. Forces that encourage change and improvement are referred to as "driving forces." The counterforces, those that work in opposition to these driving forces and discourage movement, are "restraining forces."

These restraining forces serve to assert and maintain the status quo. Often they may be negative, emotional, illogical, unconscious, and based on the desire to maintain social relationships and structures in their current state. Driving forces generally are positive, reasonable, logical, conscious, and based on economic advantage or necessity. Both sets of forces are very real and must be taken into account when implementing school restructuring.

Each of the Five Technologies represents a driving force. Implementing any one of them may bring results—for a while. But, eventually, the restraining forces will counteract any positive effects. It is like injecting a foreign substance into the human body. The body's immune system eventually counteracts and eliminates the foreign substance, and things return to normal. But when you implement all Five Technologies simultaneously, you create an environment in which you counteract most, if not all, of the restraining forces. By using the Five Technologies, you create a set of actions that help to change these restraining forces into driving ones.[3]

It would be nice to think that we could implement only one or two of the Five Technologies without paying attention to the others. But each technology builds on and contributes to the others. Educational managers who have experimented with piecemeal change have discovered the presence of uncontrollable complexity and bewildering interdependency. Actions taken in one area may have significant effects in distant parts of the system, effects obscure to the original actors. This makes it difficult for decision-makers to see the consequences of their decisions in order to learn from them.

In education, delays between when decisions are made and their impact may be very long. Often years will elapse before the totality of effect will be seen. Using the Five Technologies together can mitigate some of the most basic limits to systemic improvement. The Five Technologies work together to provide the means for certain, unambiguous, systemic, and sometimes even rapid feedback on actions taken.

Much of what you will read in the pages to come will not be "new." Many people, besides myself, have written about the individual technologies described in this book. What I say about them individually, others have said also. What most have not pointed out, though, is the mutually dependent relationship among them—the fact that *they can only work well when they are working together.* This book, I believe, for the first time, ties all of these diverse ideas and innovations together into a single cohesive strategy for change. I show how these Five Technologies are in fact crucial, interlocking components of a larger structure—a structure that is greater than the sum of its parts. As remarkable as each individual technology is, the totality represented by the Five Technologies is even greater.

HOW I CAME IN CONTACT WITH THE FIVE TECHNOLOGIES

My goal during my early years of college was to become a teacher. Both my parents were teachers and most of the adults among our family's social acquaintances during my growing-up years were teachers. And so it seems natural that I planned to become a teacher. In college, I majored in Spanish with a minor in history. I planned to teach those subjects in high school.

During the semester that I was doing my student teaching, I began to see a lot of problems at the school in which I was working. The classes were so large that it was difficult to really teach *anything.* The students did not seem motivated. Discipline was a problem. Few teaching resources were available and there were few tools for the students to use. The *modus operandi* was lecture–seatwork–homework–quizzes. Learning, for the most part, seemed to be rote, uninteresting, and temporary. What we were doing didn't seem to be working for most kids.

I realized, though, that these problems were common to most schools, and I knew that what was wrong went beyond the school, teachers, principal, or parents. The "problem" could not be blamed on the teachers or principal. They were all dedicated to their work and were doing the best they could. But something was wrong with the *system* . . .

It was about this same time that I came across West Churchman's classic book, *The Systems Approach*.[4] That book was enlightening because it helped me to begin to think about things in "systems" terms. I began to see how things related to each other, that there were often multiple causes for problems rather than just the simple cause and effect relationships I had seen before.

The book had such a profound impact on me that I decided to abandon my objective of becoming a teacher and see if there was anywhere that I could study how "systems thinking" was being applied to the problem of improving education. I discovered that Brigham Young University had an excellent program dealing with this concept. The program was called Instructional Science. It dealt with three of the five technologies: systems thinking, systems design in education, and instructional technology. So it was there I decided to do my graduate work.

This is how I came in contact with three of the Five Technologies. I eventually accepted a faculty position at the Learning Systems Institute at Florida State University. The Institute focuses exclusively on applying systems thinking to educational and adult training organizations. It was through my work at LSI that I learned about the other two technologies—quality science and change management. I recognized these as the *missing links* in school reform. Change management allows us to implement change in ways that keep it friendly to people and consumers. Quality science allows us to continuously improve the system components. I began to see how critical all five of these technologies were for true educational reform. Now, in this book, I have the opportunity to share these insights and observations with you.

NOTES

1. John Kenneth Galbraith, *The New Industrial State* (Boston: Houghton Mifflin, 1967), p. 12.
2. Peter M. Senge, *The Fifth Discipline: The Art and Practice of the Learning Organization* (New York: Doubleday, 1990), p. 6.

3. The concept of "forcefields" is derived from sociologist Kurt Lewin. The description and some of the wording used here to describe driving forces and restraining forces follows Steven R. Covey, *The Seven Habits of Highly Effective People* (New York: Simon and Schuster, 1989), p. 279.
4. C. West Churchman, *The Systems Approach* (New York: Dell, 1979).

PART I

The First Technology: Systems Thinking

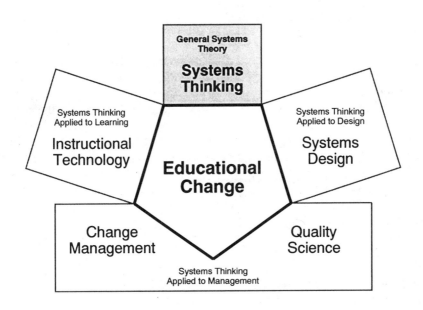

CHAPTER 2

SYSTEMS THINKING

"Man is a prisoner of his own way of thinking."
—**Stafford Beer**[1]

"The significant problems we face cannot be solved at the same level of thinking we used when we created them."
—**Albert Einstein**

As a young college student, I tended to think about problems in rather simplistic ways. The solution to the "education problem" seemed simple—hire better teachers and make classes smaller. I later came to the conclusion, as I studied systems and systems thinking, that the problem was much more complex. Most solutions address only certain parts of the problem and by themselves have little impact. I now believe that education can only be correctly perceived and understood if viewed through a properly focused "systems" lens.

This systems lens allows us to see things differently from those who look at the same problem through a different lens. Norwood Russell Hanson, in his book about the scientific method,[2] pointed out that people often see the same things but interpret them differently based on their thinking. He provides the following example: Imagine Johannes Kepler, the astronomer, watching the dawn from the top of a hill. With him is his fellow astronomer Tycho Brahe. Kepler believed that the earth revolved around the sun. Brahe believed the sun revolved around the earth. Both are watching the same event (the oncoming of dawn). However, Kepler would see the rotation of the earth slowly *revealing* a stationary sun. Brahe, on the other hand, would see the sun "rising" around a stationary earth. The truth is that people interpret what they see differently, depending on their thinking and beliefs. Systems thinking gives those who know about it "eyes to see" relationships and interconnectedness that those without systems thinking often miss.

If we look at education without a systems lens, we fall into the same trap as the proverbial blind men who were assigned the task of describing an elephant. When most people hear this story, they may fail to realize that the interesting part of the story is not that each blind man, after examining a different part of the elephant, had a completely different understanding of what he was viewing. The interesting part is that someone (the teller of the story) was able to see the *entire* elephantine system and to observe how foolish the blind men were. Systems thinking is the ladder that allows us to climb to the vantage point from which we can view the *whole* system. Systems thinking allows us to look at education by lifting the roof off, exposing the floor plan, enabling all to see the entire pattern and operation. It allows us to see how a problem in one part of the educational process may contribute to problems in other parts.

> *"For every complicated problem there is a simple solution—and it's wrong."*
>
> —H. L. Mencken

THE FIRST TECHNOLOGY

The first technology is systems thinking. Over the past two decades, systems thinking has evolved from a foundation of theoretical knowledge into a full-blown set of tools, procedures, and techniques. There are literally hundreds of books and papers coming out of the academic and business community that show how systems thinking can effectively be applied to solving problems in organizations and how it can be used to promote organizational learning. (Some of these books and resources are listed at the end of this chapter and the next chapter.) Systems thinking can be called a technology in the sense that it is forms a rational discipline for thinking about and solving human problems.

As I showed in Chapter 1, all of the Five Technologies discussed in this book can be derived from systems thinking. Systems thinking is the seed

WHAT IS AN AUTOMOBILE?

If I ask you to describe an automobile, you may immediately switch off your thinking process and simply blurt out the things you recall about your own automobile—its wheels, engine, and shape. You start by saying, 'Well, an automobile is something that has four wheels and is driven by an engine.' I (in an attempt to switch on your thinking process) ask whether a three-wheeled automobile is a possibility. You have seen one and will readily admit this change in your description, still without thinking much about the meaning of the change. I, becoming more belligerent, pursue the matter further and ask you whether a two-wheeled automobile is a possibility. You begin to look puzzled, thus indicating that your thinking has been turned on at a low voltage. I go on, being cheerfully disagreeable, and ask you whether an automobile without any wheels whatsoever is also a possibility. You become more puzzled and think not about automobiles but about silly question posers. Yet to consider the wheel-less automobile is a creative way of looking at this system we call the automobile. An automobile that can float a few feet off the surface of the earth might provide a far more comfortable ride and produce far fewer problems of traffic congestion and even of accidents. And floating automobiles may be technically feasible in the future.

The way to describe an automobile is FIRST by thinking about what it is for, about its FUNCTION, and not the list of items that make up its structure. If you begin by thinking about the function of the automobile, that is, what it is for, then you won't describe the automobile by talking about its four wheels, its engine, size, and so on. You will begin by thinking that an automobile is a mechanical means of transporting a few people from one place to another, at a certain prescribed cost. As soon as you begin to think in this manner, then your 'description' of the automobile begins to take on new and often quite radical aspects. That's the systems approach to automotive transportation.

—Adapted from C. West Churchman, **The Systems Approach** (New York: Dell, 1979), pp. 2–13.

from which the other four technologies have sprung. Without the orientation of systems thinking, we fall into the trap of seeing the Five Technologies as independent and unrelated. Systems thinking is the lens that focuses the other technologies into a powerful force for change. It allows us to explain why the Five Technologies gain their greatest power when they are working together.

Systems thinking recognizes that no action is unilateral in its impact. When we change one element of a system, we have to recognize that we will influence, in differing degrees, the other elements of that system as well. Systems thinking recognizes this and reminds us that we have to consider the dynamic, changing nature of systems.

Systems thinking is a discipline for seeing wholes. It motivates us to think about all of the elements and relationships that exist in a system, and it shows us how to structure those relationships in efficient, effective ways. In education, we are doing systems thinking when we look at education as a whole system rather than viewing it as a set of separate activities. Usually, we study education as a set of separate and distinct activities—school administration, classroom practices, curriculum, instruction, child development, assessment, policy, budgeting, etc. This is like thinking of the various systems that make up the human body (the respiratory system, the digestive system, etc.) only as separate, individual systems and not considering how they all work together and depend on each other. This fragmented approach to thinking about education results in a compartmentalized, unintegrated characterization of what education is as a whole system. Such a characterization prevents us from making improvements to the system as a whole—it limits the scope of our efforts to improving individual parts or pieces. Without systems thinking, we can only "tinker at the margin." We should not be surprised, therefore, to find that our tinkering results in small, incremental improvements. To make more dramatic progress, we must be able to understand the mutually interacting and complex dynamics of the whole system, and to strategically design interventions that deal with the system as a whole.

SYSTEMS THINKING AND EDUCATION

Classical science during the last three hundred years has sought to understand new problems by taking them apart and then seeking to understand the whole by studying the characteristics of the parts. This orientation toward reducing complex things to simple terms, while often necessary, sometimes results in oversimplification. This way of thinking often does not enable us to grasp the characteristics manifested in the interaction of parts. In the last decade, we have come to realize that this analytical, reductionist method has to be complemented with synthesis and systems thinking—approaches that are aimed at understanding larger and larger wholes.

Banathy, in his book *A Systems View of Education,* describes how the fragmented and classical mindset has been the dominant thinking that has led to fragmented efforts toward educational improvement over the last decade.

> It manifested itself in a piecemeal and incremental, disjointed, 'tinkering with the parts' approach to educational improvement. It has failed to connect and integrate the various 'solution ideas' that have emerged into an internally consistent system of solutions. It has focused on the existing system and stayed within its boundaries. At best, it stipulated changes at the margin. It has failed to recognize the complexities of current issues surrounding education as a social system; a system interacting with other societal systems, a system which is embedded in the rapidly and dynamically changing society.[3]

Systems Thinking

Without a common understanding about the systemic nature of schooling, educators will continue to institute piecemeal fixes. Piecemeal fixes are things like offering workshops to teachers, raising graduation requirements, changing labels on courses, revising the curriculum, lengthening class periods, and other such proposals. Some of these proposals and programs may be very good, but if enacted alone, without detailed analysis and consideration of how they interact with each other, they will have little, if any, positive impact.

Not only must we look at factors within the schools, we also must look at factors that affect the whole system from the outside. As one example, many observers point to the inherent absence of competitive market incentives in the structure of the government controlled public school system as the principal reason for lack of continuous improvement, modernization, and innovation. All of these factors must be given consideration as part of the systems view. Those who view reform of education only within the confines of the existing bureaucratic structure are ignoring these important external factors. These external economic factors and incentives likely have a larger impact on the performance of the system as a whole than do the myriad of smaller internal factors.

INSTANCES OF NON-SYSTEMS THINKING IN EDUCATION

Fixing Schools by Hiring a Hero Principal (or Superintendent). We can no longer rely on the traditional American solution to problems: If it doesn't work, buy a new one. For example, it is popular to think of fixing a school by getting a new principal (getting new management). Systems thinking helps us see that this is a highly oversimplified solution. We can't just bring in someone or something to fix what's wrong. We have to figure out what's wrong with the system.

Computers in the Classrooms. This is an instance of non-systems thinking because using computers for part of the instruction is only one element of the problem. Performance of the system can only be partially influenced by this solution. Much of the variance in school quality is due to larger elements and influences outside of what could possibly be influenced by a few computers in the classroom.

In-service Training. To view teacher or administrator training as the solution to the overall problem of improving education is an instance of non-system thinking because it focuses blame on poor performance by employees. Much of the variance in employee performance is due to the system in which the employees are expected to operate. Changing the behavior of the employees will not address these larger influences and may, therefore, have little effect. Also, the employees may find it difficult to apply or consistently implement new behaviors within the constraints of the current system.

Teacher Certification. Thinking that we can fix education by changing teacher certification requirements is an obvious non-systems thinking approach. Again, certification requirements only partially influence the problem, and the effects of these requirements are ambiguous and largely unknown.

School-based Governance. Switching decision-making authority from the district office to the local school level may be a good thing; however, in and of itself, it is not systems thinking. The decision-making process is only one of many factors affecting the total system. By itself, it will have little impact on student learning.

Model Schools. Creating schools that serve as models is not systems thinking because these schools are typically dependent on supplemental funding or grants that are not available to the other schools. Systems thinking would require district or state-wide structural changes and commitment of regular budget resources. The change must be accessible to all schools.

Educational Fadism. Attempting to fix education by applying the newest education fad is more non-systems thinking. We have seen fads such as instructional television, open classrooms, and new science and math curricula come and go with little effect on the outcomes of our educational system. This is largely because they were not combined into an effective systems approach.

MAJOR CONCEPTS

Here are some basic definitions of major systems thinking concepts. Each of these systems concepts has more technical definitions than those I offer here. I have tried to present them using common language and, in some cases, to provide examples from education to show how they apply.

Systems

A system is a group of parts or components working together as a functional unit. The parts of a system are subject to a common plan and serve a common purpose. Because its parts function together as a unit, a system takes on characteristics and properties of its own that are more than the sum of its parts. The quality and nature of each individual part resides in its relationship to the whole system. Each part can only be understood by looking at how it functions in relation to the whole system.

There are systems that exist naturally in the universe—the solar system, various ecosystems, the human circulatory system. Other systems (such as the education system) are created by humans to achieve a purpose. Something becomes a system when a group devises and uses procedures for making decisions, acting, and setting boundaries between the group and the rest of the world. Government, education, health care, businesses, and charitable organizations are purposeful creations. People in these systems select, organize, and carry out activities in order to attain the purposes for which the organizations were created.

Systemic

Things are part of a larger system. Things are complex. Things are connected. What you do to one part of a system will influence all the other parts. You can't eliminate one part of a system without forcing changes on all the others. Parts are dynamically interrelated and cannot be understood in isolation from the whole. Optimizing one part of the system does not necessarily prevent failure of the overall system. Systemic solutions are long-term and holistic. Contrast this with solutions that are short-term, quick-fixes, incremental, piecemeal, or disjointed—"tinkering with the parts."

"Every entity is only to be understood in terms of the way it is interwoven with the rest of the universe."
—**Alfred North Whitehead**

Systems Thinking

Systems thinking is the way we think about a problem; the way we understand the world; the way we characterize and describe a problem. To apply systems thinking to a problem means that we think about the problem as a system; we can understand and describe it as a system; we realize that there may be no single cause of the problem. This is opposed to a single cause orientation—straight "cause-and-effect" thinking—where we view problems as single, independent situations. Instead, we view problems as multiple, mutual, and interdependent. That means that one problem influences another, which, in turn, influences a third problem (problem one affects problem two, problems one and two both affect problem three, etc.). So problems have to be defined by their relationship with other problems. We acknowledge that proposed solutions will have to operate in a dynamically changing and turbulent environment.

THE "HECTIC MORNING" PROBLEM

In our family, we were experiencing a problem that recurred each morning when we were trying to get our children off to school. Because we have teenagers as well as younger children, we found ourselves dealing with different schedules and needs each morning. Our teenagers needed to be driven to early morning band practice before school; the younger children needed to be driven to school if it was raining. Also, we had made arrangements with a neighbor to take her daughter to band practice with us.

The problem was that each morning, at least one child would have trouble getting out of bed on time, there would often be some petty argument among our two teenage girls over one thing or another, our kindergartener would often dawdle and needed to be helped by a parent or older sibling, one or more children would not be able to find shoes or some article of clothing, plus each child had to prepare and pack a lunch and help make breakfast. The result was that each morning left us, the parents, emotionally drained from the last minute emergencies, contentions, and frustrations that always came up. Often someone forgot to eat breakfast or forgot her lunch or left her completed homework on the kitchen table.

When we sat down to discuss the problem, we realized that we had a situation where a variety of problems were interacting with each other. Some of the problems that were interacting were:

- children staying up too late to do homework (which was often because the child put off doing it until late evening)
- children waiting until the night before the due date to finish a report for school
- younger children dawdling in the morning
- children not having clothing or other possessions organized
- not having a specific place to keep car keys, papers to be signed by parents to go back to school, etc.
- not having a large kitchen clock that everyone could see
- not having assigned responsibilities for helping make lunches and fixing breakfast

After examining the nature of the total problem, we realized that solving any one of these individual *pieces* of the problem would not solve the overall problem (hectic morning). To solve the total problem, we would have to deal with all of the factors that influence the problem all at the same time. The result was that we held a family meeting, listed each one of the factors that led to the hectic morning problem, and made plans for addressing all of them.

Systems thinking gives us a more scientific and less reactionary mindset. It helps us respond to problems rather than just to symptoms. Without systems thinking, we run the risk of picking one symptom of a problem or situation and pressing it as hard as possible, arguing that other people are failing to see the true situation. Systems thinking helps us to stand apart, to look at what is going on, and to make a judgment about all the ramifications of the system as far as it can be seen. Systems thinking attempts, dispassionately and with a clear mind, to set down the essential ingredients in a messy situation, untangle the various factors, and set the matter straight in a scientific and objective fashion.

Systems thinking is a tool to help us "think through" a problem. It helps us avoid the "urgent need" mistake. Those who say "there is an urgent need to _____," should realize that, no matter what goes in the blank, the problem is probably more complex and needs to be considered in light of other factors that may influence it.

Open and Closed Systems

No system operates in a vacuum. All systems function by interacting, to some degree, with the things around them. *Closed systems* are those systems that have only limited and highly controlled interaction with their environment. (For example, a household thermostat: The system operates solely on one factor—temperature.) *Open systems*, on the other hand, have intensive interactions with their environment. Changes in the environment affect the operation of the system. Also, the system and what it does influence the environment, so there is a two-way effect. We would consider the current public education system to be an open system—it contributes extensively to and depends largely upon contributions from its environment. It must adapt constantly to outside changes and demands (for example, migrant children, special education students, Crack babies). The education system is expected to carry out its commitment to the environment responsibly (it must provide well-educated and skilled graduates). The environment is expected to provide the resources and support that are needed by the educational system (money, teachers, students).

Open systems have some influence on but cannot control their environment. The environment can become more demanding and even demand contradictory expectations (such as wanting the public schools to assure that all students master basic and advanced academic skills but then also requiring a large number of extra-curricular units such as "stranger danger," drug education, gang resistance education, gun safety, bicycle safety, sex education, etc.). Open systems must deal in some way with these environmental changes and demands. In an environment that changes constantly, this becomes a continuous process.

Systems are never totally open. If a system were totally open, it would have no orderly interaction with its environment. It would, in essence, be a non-system because it would lack self-definition. Openness and closedness are matter of degree.

System Boundaries

Every system has boundaries that limit and define the system itself and set it apart from other things around it. As mentioned above, a system that is totally open—no boundaries, no orderly interaction with the outside—is not a system at all. It is just sporadic activity. Setting the system boundaries is a matter of determining what types of inputs the system will accept and what outputs it will produce. In education, this involves answering such questions as: Will the school attempt to educate severely handicapped children, emotionally disturbed or violent children? Will the school attempt to deal with children's health, family, nutrition, and economic problems as well as with strictly educational issues? In education, establishing these system boundaries is always a matter of judgment. The setting of system boundaries is a crucial issue in any system. A system must regulate its input, but it can't do this too much or it will jeopardize its relevancy to its environment.

Homeostasis (Tendency to Repel Change)

Systems have a tendency to repel change the same way a human immune system rejects foreign bodies. This characteristic of systems is referred to as homeostasis—the tendency of a system to move back to a previous state of equilibrium after being disturbed by external forces. In other words, when a change is introduced into the system, the system will always work to change itself back again. This system tendency explains why it is so difficult to make serious changes in our public educational system—why many innovations and interventions do not take hold beyond their isolated pilot sites. If we want to have any kind of a major impact at all, we must plan for the multiple counteracting forces that will be exerted to change the system back to its previous state. The fourth technology (Change Management) provides tools designed to focus these potential counteracting forces toward productive changes (see Chapters 8 and 9).

Detail Complexity and Dynamic Complexity.

The terms *detail complexity* and *dynamic complexity* help explain the complex nature of systems.[4] Some systems are complex due to the high number of variables involved. This is called detail complexity. ("There are so many details involved that I can't figure out how this works!") Dynamic complexity arises when the cause and the effect are separated by time and space and are subtle. It's hard to tell what actions affect what results. ("I can't figure out what causes what!") In education, we have both detail and dynamic complexity. Detail complexity is due to the diversity of people and students involved, and also due to the myriad of tasks, methods, issues, policies, and laws that are part of the system. We have dynamic complexity because we don't know what actions directly impact which results. So we grasp at straws—endlessly implementing new programs, never really knowing if they will have a positive or negative, a small or a large effect, on the important outcomes. Dealing with dynamic complexity requires careful analysis of the relationships between actions and results. The third technology (quality science) provides the tools that help us conduct this analysis and understand the dynamic complexity involved (see Chapters 6 and 7).

Leverage

Systems thinking teaches that obvious solutions usually don't work. Obvious solutions usually create little improvement because they focus only on the surface level of the problem, and they often create less than obvious undesired longer-term effects. This does not necessarily mean that small, well-focused actions can never produce significant, enduring improvements. They can—but only if they are in the right place and considered as part of a total system. Focusing action on areas that will have the most overall impact is known as the principle of leverage. Small, well-placed actions can be like a lever and fulcrum, prying or dislodging something that imme-

---

LEVERAGE AND THE "TRIMTAB FACTOR"

On airplane wings and on the keels of racing yachts, trimtabs are small adjustable flaps that assist in balancing and steadying the motion of the craft. The principle of the trimtab also applies to a ship's rudder. In explaining the trimtab factor, [the late] Buckminster Fuller [architect, inventor, philosopher] used the image of a large oceangoing ship traveling at high speed through the water. The mass and momentum of such a vessel are enormous, and great force is required to turn its rudder and change the ship's direction. In the past, some large ships had, at the trailing edge of the main rudder, another tiny rudder—the trimtab. By exerting a small amount of pressure, one person could easily turn the ship. Thus, the trimtab factor demonstrates how the precise application of a small amount of leverage can produce a powerful effect.
—From Harold Willens, ***The Trimtab Factor: How Business Executives Can Help Solve the Nuclear Weapons Crisis*** *(New York: William Morrow & Company, 1984), p. 27.*

diately has a large impact on the total system. Tackling a difficult problem is often a matter of seeing where the high leverage lies, where a well-placed change would lead to lasting, significant improvement.[5]

Single-Loop and Double-Loop Systems

Single-loop and *double loop* are concepts that involve the feedback structures of a system.[6] Feedback is information about results, which is "fed back" to the system so that the system can modify its work if the results are not what had been intended. To understand single-loop and double-loop systems, think again of a home heating and air conditioning system. A thermostat continually monitors environmental changes, turning on a furnace to heat the room, or turning on an air conditioner to cool the room, based on a desired temperature setting. For this system to become a double-loop system, the thermostat would need to be able to adjust its goal. In other words, it would need to be able to ask itself, "Why am I set at this particular temperature?" (perhaps no one is home or perhaps only certain rooms are being used). If it could ask itself this, and could change the goal to which it has been adjusting itself, it would be a double-loop system. A double-loop system is one that can modify its underlying assumptions, policies, and objectives based on information that it receives. The second loop provides the ability to feed back information to the system so that the system can act on that data.

Double-loop systems are required whenever there are potential gaps between intentions and results. A single-loop system may efficiently and effectively achieve certain results, but these may not be the results we intended. Double-loop systems are required whenever this situation has the potential to exist. For example, our federal welfare system was designed to reduce poverty. It may, however, actually increase poverty due to the fact that it reinforces unemployment in some situations. This shows a gap between intentions and outcomes. A double-loop system is able to adjust itself if the desired outcomes are not achieved. Education systems must operate as double-loop systems, rather than as single-loop systems. They must be able to continuously evaluate the results and adjust themselves if those results are not the results that are desired.

Self-Regulation

All systems are directed toward a specific purpose. In education, that purpose is the education of students. But specific expectations change over time due to changes in society, changes in students' and families' wants and needs, changes in the workplace, and changes in the economy. Also, the inputs received by the education system change (the composition of students today is quite different from what it was 50 years ago). Thus, a system has

Systems Thinking 29

> **EXERCISES IN SYSTEMS THINKING**
>
> To sharpen your skills in systems thinking, explain why *all* of the following approaches to educational reform are examples of *non*-systems thinking. Also explain how these reforms might be made more effective through the application of systems thinking.
>
> *Reforms in subject matter* (e.g., whole language and literature approaches to language arts)
>
> *New standards* (e.g., the new standards in mathematics published by the National Council of Teachers of Mathematics)
>
> *Reforms in student assessment* (e.g., use of portfolio assessment that includes writing samples, open-ended math reasoning items, etc.)
>
> *Reforms in the professionalization of teaching* (e.g., having teachers serve as mentors to new teachers, take on new responsibilities over time, and exert more leadership through site-based decision making)
>
> *Reforms that focus on holding schools and teachers more accountable* (e.g., requiring schools to write goals and then holding them accountable for meeting those goals)
>
> *Reforms such as:* merit pay for teachers, longer school day, smaller class sizes, new school buildings

to be able to adapt itself to these kinds of changes. It might need to produce new kinds of outputs and results. How do systems make these internal changes?

There are two ways systems "self-regulate." One is *adaptation.* The other is *transformation.* Adaptation is a process for correcting errors and making modifications to bring the system in line with small changes in its environment. Adaptation only works if the changes in the environment are relatively small. Transformation must be used in the case where the system has neglected to adapt itself gradually and now faces a major gap between itself and its environment. Transformation involves rethinking the purposes, goals, processes, and outputs of the system. If a system waits too long to begin to adapt to a changing environment and increasing demands, it may need to be replaced by an entirely new system, because improving the existing system will no longer work. This may very well be the case with today's education system. Because the education system has waited too long to adapt, it may have to be entirely replaced by a new mechanism more attuned to the technology and learning methods of the modern world. Some may disagree, but it is my view that the current public education system has made such inadequate responses to a changing world and technology that its level of dysfunction is far beyond being brought into line through adaptation. Other types of systems may need to be created to replace the current one. Systems such as non-profit or for-profit schools, charter schools, electronic schools, home schools, or other educational arrangements where students and parents have more options are examples of new kinds of systems that could be created. These later examples are attempts to reform the system through transformation rather than through adaptation.

SUMMARY AND CONCLUSION

Systems thinking is the first and most important of the Five Technologies. It is the glue that holds everything together because it is through systems thinking that we are able to see how different problems relate to and sometimes even cause each other.

Problems are always embedded in other problems. Whenever we set out to solve a problem or deal with a situation, what we usually encounter is a cluster of problems. Each problem is related to every other problem, and any solution that we might propose for one problem may aggravate or interfere with others. For this reason, we have to deal with the cluster of problems as a whole. Systems thinking is the tool that helps us to look at the total problem cluster.

Systems thinking also reminds us that systems have characteristics and ways of acting that can only be understood in systems terms. An understanding of open and closed systems, homeostasis, boundaries, complexity, adaptation, and transformation helps us deal rationally with the problems of systems. Positive reform of education cannot be accomplished without an understanding of these basic systems concepts. Proposals for education reform that ignore the nature of systems will prescribe piecemeal changes that will mean little in the long run.

Finally, systems thinking explains why it is that change of the current public education system through adaptation may not be possible. Adaptation only works when relatively small changes in the existing system are sufficient. Larger changes come about only through system transformation. System transformation involves the creation of new systems of education to replace the current one. Systems thinking and the other four technologies described in the following chapters are the essential tools for designing and creating new systems of education.

FOR FURTHER READING

Russell L. Ackoff, "Towards a System of Systems Concepts," *Management Science,* 17 (1971), pp. 661–671.

Russell L. Ackoff, "Ackoff's Fables," *Systems Practice,* 2 (1989), p. 375.

Bela H. Banathy, "Systems Inquiry in Education," *Systems Practice,* 1 (June, 1988), pp. 193–212.
Bela H. Banathy, *Systems Design of Education: A Journey to Create the Future* (Englewood Cliffs, NJ: Educational Technology Publications, 1991).
Bela H. Banathy, *A Systems View of Education: Concepts and Principles for Effective Practice* (Englewood Cliffs, NJ: Educational Technology Publications, 1992).
Ludwig von Bertalanffy, "General System Theory," *General Systems,* 1 (1956), pp. 1–10.
Ludwig von Bertalanffy, *General System Theory,* rev. ed. (New York: George Braziller, 1968).
Frank Betts, "How Systems Thinking Applies to Education," *Educational Leadership* (November, 1992), pp. 38–41.
Gary D. Borich and Ron P. Jemelka, *Programs and Systems: An Evaluation Perspective* (New York: Academic Press, 1982).
Kenneth Boulding, "General Systems Theory—The Skeleton of a Science," *General Systems,* 1 (1956), p. 11.
Peter Checkland, *Systems Thinking, Systems Practice* (New York: John Wiley & Sons, 1981).
C. West Churchman, *The Systems Approach* (New York: Dell, 1979).
F. E. Emery (Ed.), *Systems Thinking* (Middlesex, England: Penguin, 1981).
Jay W. Forrester, *Principles of Systems,* 2nd ed. (Portland, OR: Productivity Press, 1968).
Michael C. Jackson, *Systems Methodology for the Management Sciences* (New York: Plenum, 1991).
Draper L. Kauffman, Jr., *An Introduction to Systems Thinking* (New York: Future Systems, Inc, 1980).
George J. Klir, *An Approach to General Systems Theory* (Princeton, NJ: Nostrand, 1968).
Colleen Lannon-Kim, "Revitalizing the Schools: A Systems Thinking Approach," *The Systems Thinker,* 2 (Cambridge, MA: Pegasus Communications, 1991).
Ellen B. Mandinach and Hugh F. Cline, *Classroom Dynamics: Implementing a Technology Based Learning Environment* (Hillsdale, NJ: Lawrence Erlbaum, 1994).
Charles M. Reigeluth, "The Imperative for Systemic Change," *Educational Technology,* 32, No. 11 (November, 1992), pp. 9–13.
Charles M. Reigeluth and Robert J. Garfinkle (Eds.), *Systemic Change in Education* (Englewood Cliffs, NJ: Educational Technology Publications, 1994).
Peter M. Senge, *The Fifth Discipline: The Art and Practice of the Learning Organization* (New York: Doubleday, 1990).
Peter M. Senge (Ed.), *The Fifth Discipline Fieldbook* (New York: Doubleday, 1994).
T. N. Warfield, "Organization and Systems Learning," *General Systems,* 27 (Louisville, KY: Society for General Systems Research, 1982).

NOTES

1. Stafford Beer, *Platform for Change* (New York: John Wiley & Sons, 1975), p. 15.
2. Norwood Russell Hanson, *Patterns of Discovery* (Cambridge: Cambridge University Press, 1958).

3. Bela H. Banathy, *A Systems View of Education: Concepts and Principles for Effective Practice* (Englewood Cliffs, NJ: Educational Technology Publications, 1992), pp. 8–9.
4. Distinguishing detail complexity from dynamic complexity is explained in Peter M. Senge, "The Leader's New Work: Building Learning Organizations," *Sloan Management Review,* 32 (Fall, 1990), p. 15.
5. The principle of leverage is another of Senge's concepts. This discussion is adapted from Peter M. Senge, "The Leader's New Work: Building Learning Organizations," *Sloan Management Review,* 32 (Fall, 1990), p. 15, available from the Center for Organizational Learning, Massachusetts Institute of Technology.
6. The concepts of double-loop and single-loop learning are explained in William Issacs and Peter M. Senge, *Overcoming Limits to Learning in Computer-Based Learning Environments,* Center for Organizational Learning, Massachusetts Institute of Technology (Cambridge, MA: MIT, undated); and also in Chris Argyris and Donald A. Schön, *Organizational Learning: A Theory and Action Perspective* (Reading, MA: Addison-Wesley, 1978).

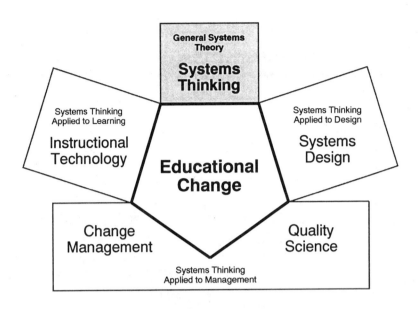

CHAPTER 3

SYSTEMS THINKING TOOLS

The principles of systems thinking have emerged from a number of different disciplines—physics, engineering, biology, psychology, and organizational behavior. Common, however, to all of these disciplines are some basic thinking tools. These tools are designed to help us think about the world around us in a more complete way. They help us think about system-wide interrelationships and connections between things.

SYSTEM DIAGRAM

One of the most basic tools for systems thinking are system diagrams. These diagrams help us see the circles of influence that exist in any action or pattern. Peter Senge, in his book *The Fifth Discipline* (Doubleday, 1990), provides several simple examples of system diagrams. He starts with diagramming the simple activity of filling a glass of water. When we fill a glass of water, we are actually operating a very simple water regulation system. We watch the water level rise in the glass as we fill it. We continuously monitor the "gap" between the current water level and the desired water level. As the water level rises, we regulate the input by adjusting the faucet position to slow or speed up the rise of the water level, and we turn off the flow when the glass is full.

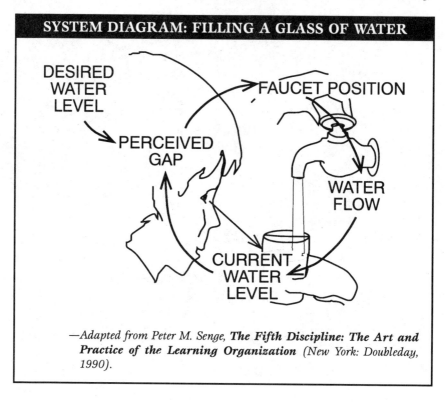

—Adapted from Peter M. Senge, **The Fifth Discipline: The Art and Practice of the Learning Organization** (New York: Doubleday, 1990).

This simple water regulation system can be diagrammed as shown in the insert "System Diagram: Filling a Glass of Water." All of the factors in the system (desired water level, the glass's current water level, the gap between the two, the faucet position, the water flow) are indicated and arrows show the relationship between these factors.

System diagrams are always composed of circles of influence rather than of straight lines. This is because patterns tend to repeat themselves, making a situation better or worse. The arrows in a system diagram represent influence on another element. You read a system diagram by saying the word "influences" each time there is an arrow represented. For example, you read the water faucet diagram, by saying:

Faucet position *influences* water flow;
Water flow *influences* current water level;
Current water level *influences* perceived gap;
Desired water level *influences* perceived gap;
Perceived gap *influences* faucet position;
and so on around the circle continuously.

Systems Thinking Tools

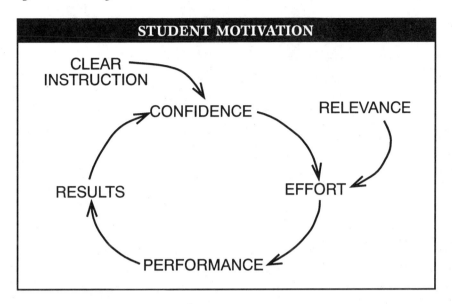

The important thing to perceive about a system diagram is that it captures the whole story, not just a part of it. It shows everything involved in the process and how one action is influenced by the other actions in the process. System diagrams help us get away from thinking that something has only one cause. We can see, for example, from the glass of water diagram, that the faucet position causes the water level to rise. But it is also true that the water level controls the rate of flow from the faucet. The system diagram of the glass of water example shows the amount of learning that can take place by trying to diagram something in this way.

A very constructive exercise in systems thinking is to form groups that are given the task of constructing a system diagram of a situation or problem they are dealing with. The problem might be student motivation, low math scores, frequency of discipline problems, customer dissatisfaction, or other types of problems. Constructing a system diagram of any of these problems forces the group to think about the major factors that act to influence the problem. By thinking in this way, they will be much more likely to generate credible solution tactics than if they use a single-cause mentality. The insert "Student Motivation" shows a system diagram constructed by a group of teachers showing influences on student motivation.

MODELING TOOLS

Systems thinking is about understanding how things work and about making predictions. Modeling tools or modeling diagrams show the chang-

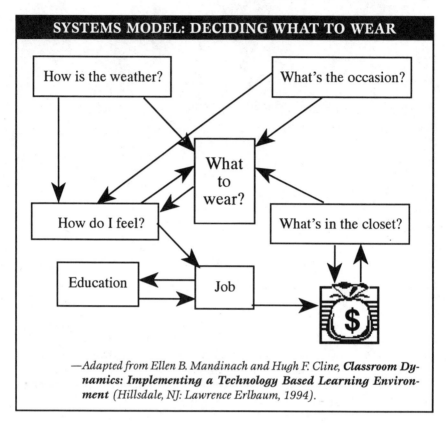

—Adapted from Ellen B. Mandinach and Hugh F. Cline, **Classroom Dynamics: Implementing a Technology Based Learning Environment** (Hillsdale, NJ: Lawrence Erlbaum, 1994).

ing relationships among the parts of any process or activity. Drawing a model of all of these parts helps us understand the nature of the system we are examining. A good way to learn systems modeling is to begin with a simple exercise that is familiar, such as getting dressed in the morning. When people decide what clothing they will wear each day, there are a variety of factors involved in the decision (the weather, the occasion, the person's mood, the type of clothing available, etc.).

The diagram in the insert "Systems Model: Deciding What to Wear" shows this getting dressed problem as a systems model. In the diagram, the two-directional arrows connecting some of the factors indicate that the causal connection could go in both directions. For example, a person's clothing budget might influence what's in the closet and vice-versa. What one wears, however, will not influence the weather, so, in that case, the arrows are one-directional. As you can see, the systems model depicts a much more complicated picture of the daily task of deciding what to wear. Using a modeling tool such as this allows us to see that a problem we consider simple, may, in fact, involve many more factors than we previously thought. Drawing a systems model thereby increases our appreciation for the complexity of a problem and allows us to see more clearly why a problem may exist and how it might be influenced. Systems models show repeating patterns and relationships among parts. They are useful tools in helping us understand the true complexity of a problem or situation.

STELLA II

As the interest in using systems thinking continues to grow, so does the need for new tools to support those efforts. One of the most widely-used systems thinking tools is STELLA.[1] STELLA is a software program that runs on the Macintosh computer. It is designed to help people build system diagrams and systems models. It is a tool kit for building, simulating, and modifying visual representations of systems. To use STELLA, you begin with a screen that contains various building blocks, tools, and objects. These are located across the top border of the screen. On the left margin of the screen, toward the top, are the navigation arrows and a toggle switch

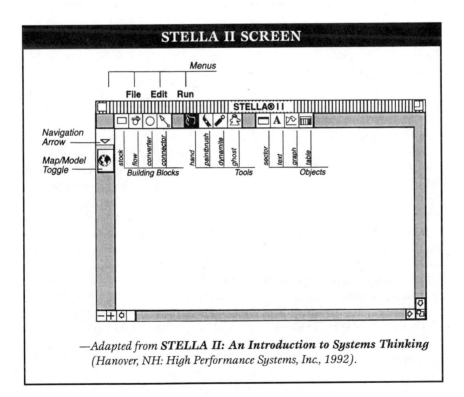

*—Adapted from **STELLA II: An Introduction to Systems Thinking** (Hanover, NH: High Performance Systems, Inc., 1992).*

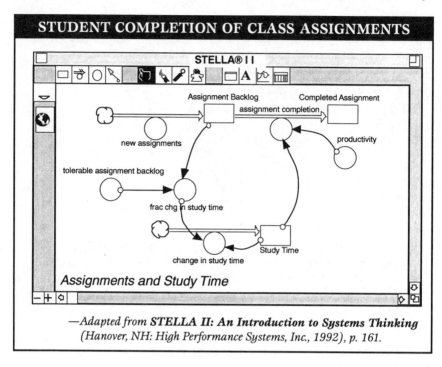

—Adapted from **STELLA II: An Introduction to Systems Thinking** (Hanover, NH: High Performance Systems, Inc., 1992), p. 161.

that alternates your view back and forth between STELLA's mapping mode and modeling mode.

To diagram a system, you pull down the various building blocks from the top border and place them into the diagram. You can use stock building blocks, flow arrows, connectors, sector frames, and other such tools. The STELLA manual describes all of these tools and how to use them.

A simple example of a systems model is given in the insert "Student Completion of Class Assignments." This model depicts the problem of student completion of class assignments. The top row of the diagram simply says that: *new assignments* build the *assignment backlog*. The student depletes this backlog through the assignment completion flow. The bottom part of the diagram shows that assignments are completed as the student spends *study time*. Notice the two circles between assignments backlog and *study time*. One circle is labeled "frac chg in study time" (abbreviation for "fraction change in study time") and another "change in study time." When the assignment backlog rises, the student increases study time, thus increasing the flow of assignment completion. As the assignment backlog falls, study time is cut. The adjustment of hours thus represents a decision on the part of the student. This mechanism works to keep the assignment backlog at a tolerable level for the individual.

Systems Thinking Tools

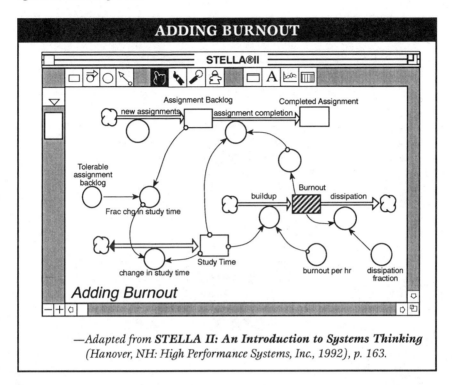

—Adapted from ***STELLA II: An Introduction to Systems Thinking***
(Hanover, NH: High Performance Systems, Inc., 1992), p. 163.

The insert "Adding Burnout" brings student burnout to the model. The shaded box (center right portion of the screen) represents burnout. The arrows to the left and right of the shaded box indicate that burnout builds up as study time increases and dissipates with time. Increased burnout impacts the student's productivity in completing assignments. The greater the degree of burnout, the lower the productivity in getting assignments done.

Stella II allows you to add numerical indices to your model and use mathematical functions to depict the effect of changes in the model on particular outcomes. You can then predict the effect of an increased level of assignments on study time and on burnout, and produce graphs that show the results. The next insert (see page 40) shows the effect of burnout on a student who waits until the 11th hour to study.

OTHER SOURCES FOR SYSTEMS THINKING TOOLS

The *Newsletter for Systems Thinking in Higher Education* is published by Cambridge Decision Dynamics of Cambridge, Massachusetts. The news-

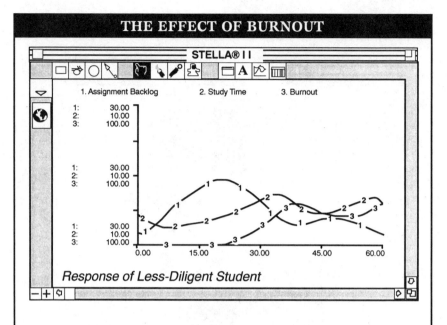

Note that burnout affects both study time and assignment backlog and is also affected by them. As the assignment backlog starts to increase so does the student's study time in this example. Burnout starts to build up once the student starts to spend overtime hours in studying. Burnout grows rapidly when study time is at its highest level. Although burnout remains high, this particular student continues to function.

—Adapted from *STELLA II: An Introduction to Systems Thinking* (Hanover, NH: High Performance Systems, Inc., 1992).

letter features short articles about applying systems thinking, case studies, methods and tools, interviews with experts, and book reviews. It also provides information about events and conferences on systems thinking.[2] Pegasus Communications, also out of Cambridge, Massachusetts, publishes *The Systems Thinker*. This newsletter lists systems thinking resources and contains short case studies and articles.[3] The Creative Learning Exchange (an organization that encourages teaching students in kindergarten through twelfth grade about systems thinking) publishes a newsletter to its network of teachers and administrators who are interested in the application of systems education. The newsletter reports on various activities and projects in elementary and secondary schools around the country. The newsletter also contains ideas for introducing systems education into different subject-matter classes and provides teaching ideas and examples of student work.[4]

FOR FURTHER READING

Daniel H. Kim, *Systems Thinking Tools, Toolbox Reprint Series* (Cambridge, MA: Pegasus Communications, 1994).

Ellen B. Mandinach and Hugh F. Cline, *Classroom Dynamics: Implementing a Technology Based Learning Environment* (Hillsdale, NJ: Lawrence Erlbaum, 1994).

John D. W. Morecroft and John Sterman (Eds.), *Modeling for Learning Organizations* (Portland, OR: Productivity Press, 1994).

Barry Richmond and S. Peterson, *STELLA II* [Computer program] (Lyme, NH: High Performance Systems).

S. B. Robinson, "STELLA: Modeling and Simulation Software for Use with the Macintosh," *Byte* (December, 1986), pp. 277–278.

Peter M. Senge, *The Fifth Discipline: The Art and Practice of the Learning Organization* (New York: Doubleday, 1990). See Chapter 5 for a discussion of systems diagrams.

NOTES

1. STELLA is an acronym for Structural Thinking Experiential Learning Laboratory with Animation. The STELLA computer program is available from High Performance Systems, Inc., 45 Lyme Road, Hanover NH 03755. Tel. 603-643-9636. Fax 603-643-9502.
2. To subscribe to the *Newsletter for Systems Thinking in Higher Education,* write to Cambridge Decision Dynamics, Suite 125, 26 Landsdowne St., University Park at MIT, Cambridge, MA 02139. Tel. 617-252-0909. Fax 617-252-0889. E-mail DDDynamics@aol.com.
3. Contact Pegasus Communications, Inc., P.O. Box 120, Kendall Square, Cambridge, MA 02142-0001. Tel. 617-576-1231. Fax 617-876-6426.
4. To subscribe to the Creative Learning Exchange newsletter or to join the Creative Learning Exchange, write to The Creative Learning Exchange, 1 Keefe Road, Acton, MA 01720. Tel. 508-287-0070. Fax 508-287-0080. The Creative Learning Exchange has a home page on the World Wide Web. The address for the Creative Learning Exchange home page is htpp://sysdyn.mit.edu.

PART II
The Second Technology: Systems Design

Design is _not_ repair.
Design is _not_ improvement.
Design is _new_.

Design is erasing the assumptions of the current system and replacing them with assumptions that more closely represent what we know...

Design directs thought and action toward reaching a goal.

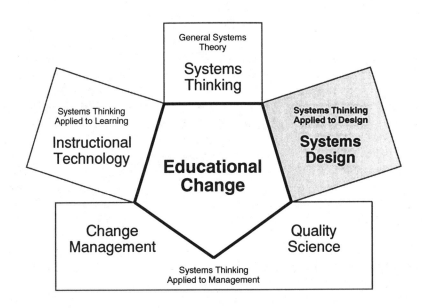

CHAPTER 4

SYSTEMS DESIGN

"Where we are going there are no roads."
—Bela Banathy

Systems design is the application of systems thinking to the process of design. When we think of design, we typically think of designing products or commodities, houses, or office buildings. But we also design systems. When we design systems, we are designing structures in which people, processes, tools, and machines combine to accomplish a predetermined purpose. Systems design includes planning the use of time, money, and human resources; designing the facilities, the processes, the flow of materials; and planning all other activities that work together to accomplish the purpose for which the system is intended.

Designing a new system is different from repairing or improving an existing system. If you think you are in the business of repairing something, then you accept its current structure and say: "All right, I'll accept that it ought to be sort of like it already is; but let me just do two little things to make it a bit better." In repairing something, we begin with the structure that already exists and go on from there. Design is different because, when we design, we start with a *vision of the ideal system* and we create totally new kinds of organizational structures to accomplish the mission for which the new system is needed.

Designing a new educational system deals with the business of replacing the current system of schooling with something potentially quite different—not with repairing what now exists. Therefore, we say: "Let's erase the assumptions of the current system and replace them with assumptions that more closely represent what we know about about how children learn, about instructional methods, about quality, incentives, and about modern management methods." If we think of schools as they are organized, administered, and funded today, we restrict our thinking to what already exists. We will likely consider only those ideas that can be done in the confines of today's schools and bureaucratic systems. However, if we think about education as the creation of educational services that are provided to students and families, a whole new range of possibilities opens to us. We now can be creative and think of ways to arrange resources, methods, people, buildings, and technological devices to provide education. Schools within schools become a possibility. Learning at home or at other locations in the community may be possible. Charter schools, scholarships to private schools, or outdoor schools can be considered. We could design a curriculum that is available 24 hours a day via electronic media. School might be a place devoted more to peer tutoring, small group work, cooperative learning among students working on research projects, and high tech learning labs as opposed to classrooms and straight teacher presentation. As soon as we begin to think about what education is *for,* instead of how it is typically done, our thinking begins to take on new and often radical forms.

Design belongs to the category of activities known as *teleological.* The word comes from the Greek *telos,* meaning end, and conveys the idea that things can be directed toward an end or shaped by a purpose. In this respect, design is similar to planning, building, and engineering. All of these activities are teleological in nature because they direct thought and action toward reaching a goal. Design is thinking about and creating alternative solutions in order to figure out which alternative will most effectively and efficiently lead to the desired goal or outcome. It is going where there are no roads, creating alternatives that have not yet been thought of.

As an illustration, suppose our goal is to "solve the problems of education." We might begin by analyzing the existing system, identifying instances or situations where the system does not now work effectively, and then propose solutions for those situations. Do you see that the approach we have just described is *not* systems design? It is not design because we began by using the existing system as the starting point. In systems design, we begin with a careful formulation of the goals and mission of education and then work backwards from there to create alternative visions of systems that might be designed to perform the functions necessary to accomplish that purpose.

Starting with what now exists and trying to fix it usually leads to piecemeal, disconnected, and fragmented solutions. These types of solutions are appropriate if the system is not too far out of sync with its mission. But when a system is out of sync with its environment, when it reaches the point at which it is no longer capable of achieving its purpose, we need to design a different system. Small improvements in the current system are easy to perceive but will not bring about the overall results that are needed. Design is hard to perceive, but it is design that will allow the large payoffs in results.

WE CANNOT BE CONSTRAINED IN OUR THINKING

Design requires that we are not constrained in our thinking by current practice or by limitations that may not be relevant in the future. Our thinking is often so influenced by current practice that it becomes difficult to conceive of models or approaches that are not close copies of the current way of doing things.

Most recommendations for restructuring education that we read about in the reform literature are attempts to fix, modify, or adjust the current model of education by adding supplemental programs, lengthening the school day, or increasing or decreasing certification requirements. None of these, one should note, change the fundamental operational model of schooling. Cosmetic change—making things look better on the surface—will not work, if the real need involves resdesign of the whole enterprise.

In his article "The Great School Reform Hoax," George Leonard commented on the lack of fundamental changes in schooling by observing that:

> Over the past hundred years almost every aspect of our national life—industry, transportation, communication, computation, entertainment—has changed almost beyond recognition. Our schools remain essentially the same. And now, in the space age, the reformers are offering the nation an educational horse and buggy. They would improve the buggy, keep the passengers in it longer, and pay the driver more. But it would still be a horse and buggy.[1]

If we are going to design successfully, we cannot limit our thinking to what can easily be done within the constraints of the current model. Rather than improving a horse and buggy, we need to design and build a new mode of space-age transportation.

Individual, piecemeal changes in the current model of schooling are not very efficient or effective ways to improve the system, from the overall point of view. Consider the "efficiency expert" in an industrial firm or school who walks around the factory and discovers that certain processes are not working very well. He recommends changes in these processes that will fix the problems. His recommendations may make perfectly good sense

> ### A WAREHOUSE FULL OF PARTS THAT DON'T FIT INTO A WHOLE
>
> The outcome of years of educational research and development efforts is nicely displayed and fills whole libraries. There is a poignant metaphor to this: a warehouse full of many vehicle parts that do not fit into a whole. The blueprint for integrating the parts does not exist. The many educational improvement programs and products do not 'map' into the system we call education in an internally consistent way.
> —Bela H. Banathy, "Systems Inquiry in Education," **Systems Practice**, 1, No. 2 (1988), p. 197.

within the narrow view of that factory. But systems thinking argues against piecemeal change because it is always thought of in relation to a small segment of the total organization. Introducing piecemeal changes by themselves may do the very opposite of what was intended. In fact, in many instances, the changes may increase the total system cost or decrease total system effectiveness. This is because piecemeal changes may simply allow us to do the wrong thing faster or better—and that's no help at all.

NEW TYPES OF ORGANIZATIONS

To achieve significant increases in student productivity, it will be necessary to design totally new types of organizations for learning—organizations that depart significantly from the existing school structure. Why will these new organizations be different from the existing structure? Because the current structure of schools was designed before the modern management and instructional methods that we know about today were invented. The past 40 years have seen major developments in thinking about organizational structures, operations, and management methods. The rigid, hierarchical organizational structures that predominated in the industrial age corporations of the past are being replaced by systems models. These models rely on systems thinking to pursue the goal of continuous improvement and consistent quality. These modern management methods use statistical analysis to isolate problem points in the process. They also use teamwork, expert systems, electronic performance support systems, and so forth to assure that everything done by the organization is done right—every time. Although these ideas and methods have revolutionized business and industry in the U.S. and in the rest of the industrialized world during the past 20 or so years, they are just now beginning to have an influence in education.

Systems Design

A second reason new organizational structures will be different from the current ones is that the current organization of schools is designed around teachers and teaching rather than learners and learning. To describe the situation more specifically, what is taught is based upon the topics to be covered and developed by the instructor, where the goal is content coverage. Teachers generally select materials based on the needs of the average student. Teachers often use a lecture, discussion, and textbook format as their primary teaching strategy. The system presents a fixed amount of content in a group-based delivery. Some students effectively learn the material, others simply flunk out. In an organization centered around the learner and learning, students would be the principal actors, with the teacher providing guidance and helping the students access the information and learning materials needed. Each student would have his or her own individual learning plan and goals as decided by the teacher together with the student's parents, with input from the student as well. (This is already being done very successfully in special education programs with handicapped or learning disabled children and with gifted students.) Different types of learning arrangements would be available to individual students depending on their needs. Options for learning would include self-directed learning, peer tutoring, small group instruction, cooperative groups, computer-assisted instruction, self-tutorials, and other methods. The mastery of prerequisite capabilities would be the only requirement for starting to work on a new learning task. Students would not need to wait for the entire group to move up, but could move on to the next step as soon as they were ready.

A third reason new organizational structures will be different from the current ones is that today, vastly more capable technologies are available that permit large improvements in instructional procedures, information handling, and learning methods. It is now possible, through the use of computer-based learning systems, for a student to find out precisely what his or her skills and learning needs are and then to be directed immediately to the learning resources that are most appropriate. Such learning methods make the whole idea of "flunking out" obsolete. In addition, new technologies such as computer simulations, computer-supported learning environments that allow students to interact with each other, with teachers, and with outside experts over a computer network, and multimedia learning systems are available and could be used to dramatically improve the nature of individual learning and its effectiveness and efficiency. Chapters 10 and 11 discuss these new technologies and the integral role they will play in the future of education.

In addition, changes in society call for new kinds of organizations. The American education system has evolved along with society in a co-evolutionary process. During the agricultural era of our society, education of children was accomplished mainly by itinerant tutors hired by the family or in a one room schoolhouse. During the industrial era that followed, we saw the invention of the comprehensive high school with its 50 minute time periods, its ringing of bells to signal when to move on to the next class, and instruction delivered to large groups. Now, our education system must evolve again as society moves away from an industrial era into an information and communication age.

The insert "General Characteristics of Industrial Age and Information Age" compares some basic characteristics of the past industrial age with

GENERAL CHARACTERISTICS OF INDUSTRIAL AGE AND INFORMATION AGE

	Industrial Age	Information Age
Purpose and Mode	Processes organized around mechanical energy for material production.	Processes organized around intellectual technology for information and knowledge development.
Power Base	Extension of our physical powers by use of machines.	Extension of our cognitive powers by high technology.
Dominant Paradigm	Newtonian classical science, deterministic, reductionist, single causality, organized simplicity.	Systems thinking, emerging mutual causality, dynamic complexity, ecological orientation.
Technologies	Manufacturing, fabricating, engineering, etc.	Gathering, organizing, storing information; communicating, networking, and systems planning and design.
Principal Commodities	Energy, raw and processed materials, machines and manufactured products.	Information and knowledge are the key products. Human, professional, and technical services are principal commodities.
Consumption patterns	Bigger and more.	Smaller and more efficient.
Organizational Characteristics	Stability and quantitative growth.	Integration, synergy, change, and flexibility.

—Adapted from Charles M. Reigeluth, Bela H. Banathy, and Jeannette R. Olson (Eds.), **Comprehensive Systems Design: A New Educational Technology**, NATO ASI Series F Volume, ARW 900426 (December, 1990), p. 10

Systems Design

those of the emerging information and communication age. These characteristics are only a sample of the massive changes that have occurred in our society over the past few decades. The characteristics of the past industrial age are fundamentally different from those of the information and communication age. The current educational system, with its roots in the industrial age framework, cannot be "improved" or "restructured" to match the characteristics of the new information age. If our standard of living is to continue to improve, we must make another fundamental change in our education system.

FEAR OF SYSTEMS DESIGN

Aren't systems ways of losing personal control by turning control over to some government or "big brother" organization? Can't people acting alone, outside of a system or organization, perform better than they can within the constraints of a system?

These are interesting questions because they reveal one of the common misunderstandings about systems. The task of systems design is to create organizations of people, processes, and equipment that will perform tasks better than a single person acting alone. Consider the example of a transportation system. A person can walk approximately four miles in an hour. But if human intelligence and modern technologies are combined to create a modern railroad or air transportation system, humans can now travel several hundred or thousand times as fast as the lone person. Our system for building houses provides us better shelter from our environment than does our own skin. Our medical and health care system prevents disease far better than a single person can on his own. Different systems of education provide resources and opportunities for learning that go beyond what any person can do individually. Systems produce these superior results simply because they coordinate and synthesize the efforts of many people with divergent skills and focus these skills toward a single task. The results of this coordination and unified effort are far superior to those that a single individual can hope to produce.

Systems design is a powerful tool—a tool that can increase our individual capabilities. Like any powerful tool, it can be used for good or ill. It can be used to enhance personal freedom and dignity—or to do just the opposite. It can be used by any person or group that wants to be effective in accomplishing its purpose. The reaction of many people to the idea of systems design is: "It sounds good. It also sounds dangerous." It sounds good to say that the whole system is being considered, but it scares us because so much mischief can be created under the guise of serving the whole. As with any tool, the values, personal ethics, purposes, and objectives of those who use systems design will influence the outcome. System design should be feared greatly when it is in the hands those who would use it to manipulate or control people. In the hands of others, it can (as we have shown in the preceding paragraph) do a great deal of good.

Isn't systems design the same thing as socialism, big government, and central control? The answer is a resounding NO! First of all, systems design can be used to design any kind of system. It can be used to design a socialist economic system or one based on free enterprise, individual initiative, and competition. It can be used to design a public education system, a system of private schools, or an education system based on electronic telecommunications—where there are no school buildings at all. It can be used to design management systems, manufacturing systems, transportation systems, and so on. When systems design is used, it will always reflect the values and purposes of those who participate. The design should result in a system that reflects the true values and goals of those for whom the system is intended.

Businesses pay a lot of attention to customer input when they design new products or services because our free market system allows people to choose what products or services to buy. The free market system helps assure that the products and services that businesses produce are those that are most valued by the people they serve. This keeps the system responsive to the individual. A large part of the problem with the public education system is that there is no direct accountability to the consumer. Public schooling today is basically a bureaucratic monopoly. For this reason it does not always act in the interest of consumers. If a different system of schooling were created, one that *did* allow direct accountability to the consumers, the system would be much more responsive and would better serve people's needs.

WHY DESIGN IS NECESSARY

Along the S^1 Curve

In Chapter 1, systems thinking was described as a lens that gave those who used it "eyes to see." By using systems thinking to look at educational reform, we can see why the various improvement and reform efforts of the last decade fell far short of expectations. Extending the school day and year, making classes smaller, offering tougher courses, putting computers in classrooms, installing computer labs, or funding model schools have produced little improvement. This is because most of these improvements and reforms are only slight modifications of past practices. Our view of the process of education as being bound to past traditions limits our thinking. This outmoded perception hinders the development of a new mind-set—new ways of thinking about organizations that could be designed to promote and foster learning.

Systems Design

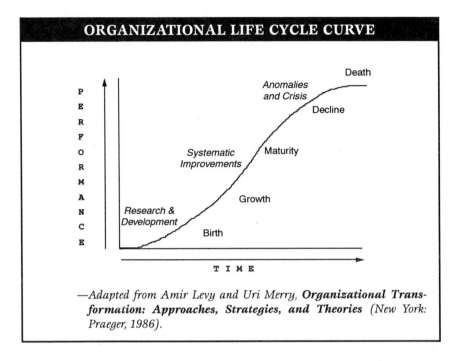

—Adapted from Amir Levy and Uri Merry, *Organizational Transformation: Approaches, Strategies, and Theories* (New York: Praeger, 1986).

We seem to be stuck on an S^1 curve. What is an S^1 curve? Those who study organizations have shown that most organizations tend to follow a similar pattern of growth and maturation. The pattern is referred to as an organization's "life cycle" and is described as an upward curve in the shape of the letter "S."[2]

The S-curve depicts the birth, growth, maturation, decline, and death of organizations. An organization, in its early stages of development, generally experiences a period of initially slow progress while it defines it mission and refines its operations. This is followed by a period of rapid growth with explosive improvements in productivity, speed, efficiency, or effectiveness. Gradually, the organization reaches maturity, at which point further improvements are difficult. Large amounts of effort and financial resources are needed to produce even small improvements once maturity is reached. This is why Harrison Owen says of organizations, "The more perfect an organization runs, the closer it is to its end."[3] If the organization persists only in refining its old techniques, if it continues to make the same once successful responses in the face of new and changing challenges, it will experience anomalies (abnormal or irregular problems) and crises, leading to a period of decline and eventually death (or at least obsolescence). The orga-

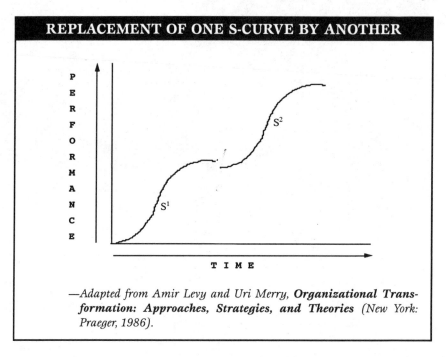

—Adapted from Amir Levy and Uri Merry, **Organizational Transformation: Approaches, Strategies, and Theories** (New York: Praeger, 1986).

nization "dies" at its level of peak performance because it gets better and better at doing something that isn't needed.

What happens when an organization has reached the top of the S^1-curve? In such situations, the organization must make a transition to a new S-curve, called "S^2." To make this transition, the organization must recognize that further improvements on their existing processes or programs will produce only small improvements. To make large improvement, it must make fundamental changes. Organizations that resist making such changes ultimately do not survive.[4]

S^1 to S^2 transitions are what ship builders had to do when they converted from building sailing ships to steam ships. It is what U.S. auto companies had to do to compete with their overseas competitors. It is what manufacturing has had to do time and time again over the centuries to convert from the master craftsman hand manufacturing methods of the early 1800s to the mass production methods of the 1850s, and finally to the modern computer-aided design manufacturing systems of today.

The birth, growth, maturity, and decline of the education system in America has closely followed the pattern of the S^1-curve. It is now stuck at the top of an S^1-curve that began when education converted from the one-room school house to the industrial model of education that we have today—with its Carnegie units, ringing bells, and 50 minute class periods. And it has not made the transition to a new curve. Moving to S^2 is essential if schools are to meet the educational needs of children in the future.[5]

THE "INSIDE-VIEW" VERSUS "OUTSIDE-VIEW"

Educational reform efforts over the past two decades have dwelt on the problems and inadequacies of the current system of education. Reform-minded individuals and groups have often viewed the system from within (rather than looking at it from without, through the lens of systems thinking). This "inside" limited view has led them to ask such questions as:

- What is wrong with the system?
- How can we improve it?
- How can we improve teachers?
- How can we focus more on the basics?
- How can we get more parent and community involvement?

These are "inside-view" questions motivated by "inside-view" thinking. These types of questions are completely legitimate to ask when a system is meeting only minor challenges. But in times of turbulence, when a system is functioning far below desired levels, and where there is a large gap between the results that are being achieved and those desired—it is time to take an outside view and ask a different set of questions:

- What are the desires of of our customers (parents, students, employers, colleges, etc.)?
- What are the educational implications of those desires?
- How can we rethink education to be better able to meet customers' wants and needs?
- What new images emerge from all of the above that might guide the design of completely new systems of learning?
- How can we use the principles and strategies known as systems design to create and implement a new system?[6]

These outside-view questions are quite different from the inside-view questions reformers have generally been asking. Inside-view questions focus on improving the system we already have; outside-view questions shift our focus from improving the system we have now to exploring the possibility of a new system. If we are going to meet the future challenges of an information and technological age, we will have to stop asking inside-view questions (and stop trying to improve the current system). Instead, we will need to use systems design to create new and more effective systems of learning and teaching.

DESIGNING FOR HUMAN PERFORMANCE

Part of systems design includes designing job roles that can be performed well by the people who will work in the system. An organization cannot depend on the availability of an endless supply of super-hero performers. The goal is to build organizational structures in which all normally skilled employees achieve excellent results. This can be done by effective job role design, effective training, and by the use of productivity and performance enhancing tools such as electronic reference systems, computer automated forms and procedures, and job performance aids such as reference sheets, job aids, and checklists.

In education, we cannot think that we can solve our problems by simply hiring the "hero" principal or the "stellar" teacher. If we think this way, we are missing the point behind the idea of designing for human performance. Only in jobs for which there are many, many more applicants than positions (such as professional sports teams, astronauts, symphony orchestras, etc.) can we depend on exceptional and unique performance within the organization. In most organizations, the task is to design job roles that can be performed well by those who are currently employed and those who we can reasonably expect to hire. If the job requires a hero or stellar performer, then we must redesign the job or provide support tools, to make success possible for those we actually can employ.

Simply getting people to work harder or better doesn't do it. If people's efforts are to bear fruit, those efforts must be supported by structural changes and incentives in the organization. Part of systems design is to design incentive structures that are consistent with actual expectations and with the organization's goals and objectives. The goal is to have the employees' monetary incentives, group status, professional advancement, and community perception all aligned with actual expectations. If there is a conflict between the stated expectations and the incentives, operations break down, an "us and them" mentality develops, and the energy of the organization is diverted toward resolving internal conflicts instead of being focused on accomplishing the organization's mission.

W. Edwards Deming, the father of the modern quality movement, estimates that 85% of the problems that affect quality reside in the organization's structure, *not* in individual people's performance. In other words, only about 15% of the problems or inefficiencies that occur in an organization are under an employee's control. Most of the problems are caused by the *system*. For example, teachers in an elementary school cannot do a top quality job when there are constant interruptions on the intercom, when they are working with faulty instructional materials, when they lack resources, or when they don't have an effective and convenient way to communicate with individual students and parents. We must, therefore, design a system that adequately supports the successful job functions of the people

Systems Design

who work in the system. We design for human performance by looking at all the processes that surround and support each employee's job role and then designing those job roles so that they can be carried out successfully by *each employee!*

Some might ask if such an analytical approach to job role design doesn't lead to mindless automation and result in loss of human initiative and creativity. On the contrary, rich experience and judgment, leadership, brilliance, and entrepreneurship are valued in any organization. But if the success of the organization depends on each employee, each administrator, and each teacher being a super performer, the law of averages will win out. The jobs must be able to be performed successfully by all employees, most of whom are somewhere near the midpoint between super performer and low performer.

THE DESIGN PROCESS

The design process can be described in 10 steps. These 10 design steps are taken from *Comprehensive Systems Design: A New Educational Technology,* by Charles Reigeluth, Bela Banathy, and Jeannette Olson. I am greatly indebted to their excellent description of these steps and the design process.[7]

1. Diagnose and Describe the Design Goal

Describe the mission that the new system is to accomplish and the kind of a system we want to design. Our goal is to create a system (an organized structure that uses people, processes, and equipment) to accomplish a predetermined purpose. What is that purpose and what type of a system will be needed to achieve it?

2. Clarify the Reason for Engaging in Design

We engage in design when we agree that existing structures are not capable of meeting a perceived need. What are the needs that the current system is unable to meet and why do we need to design a new system? If slight modifications to the current system will allow the system to meet the needs, we do not need to design a new one. We engage in design when we need a new solution.

3. Establish the Boundaries of the Design

Boundaries are drawn to mark and set apart the system within its environment. To determine these boundaries, we identify the range of activities and goals for the system we will design. We want to draw the boundaries narrow enough so that the system will have the organizational capacity, the human capability, and financial resources sufficient to enable it to achieve its stated mission and purpose. We also examine the environment that lies immediately outside of the boundaries to understand how the system will interact with that environment to acquire the various resources needed for the operation of the system and to see the different ways that the environment depends on the success of the system.

4. Formulate Core Ideas, Values, and an Image of the Future System that Will Guide the Design

The basic core ideas and values that will guide the design must be agreed upon in advance. These might be principles like the following: "the design will focus on the learner as the key element"; "the design will promote innovation and improvement, and focus on quality"; "the design will be possible to implement within existing budget constraints"; or other such principles. We should decide if we value system efficiency, effectiveness, or cost most highly; whether we want to promote individual initiative, privacy, individualism, parental choice and responsibility, and market competition, and we should discuss and prioritize other such values. We also create an image or description of the future system. This image might be in the form of diagrams showing students' progress through the system, a "day-in-the-life" in the redesigned system from the students', parents', and teachers' viewpoints, or descriptions of particular processes or activities as they would occur in the new system.

5. Define Expectations, Aspirations, Purposes, and Requirements of the System to Be Designed

The design goals were stated in Step 1. Here we state in much more detail what we want the system to do; who the clients of the system are (students, parents, taxpayers, business community, etc.); the educational services that should be offered to them; the characteristics those services should have; where, when, and how those services should be provided; what kinds of responsibilities the system should have toward its clients, the community, and the larger society; how the system should relate to the various levels of government; and the relationships the system will have with other systems (such as the public health, the legal system, and state child welfare systems). In formulating these descriptions, we should consider alternatives, evaluate each alternative, and select the alternative that is best aligned with the core ideas and values. These details constitute the system requirements and will be used as the basis for designing the system.

Systems Design

6. Create and Evaluate Alternative Representations of the Future System

In the design process, we want to create several alternative designs, not just one. These alternative designs can then be submitted to different audiences with diverse viewpoints for reaction and input. It is much cheaper to create and evaluate these alternative designs than it is to build and test alternative systems. Designing and evaluating several alternative designs can produce very high payoffs in terms of improving the final design. Looking at a variety of alternatives forces us to address more design issues than a single alternative would address.

7. Establish Criteria by Which to Evaluate the Alternatives

Before submitting the alternatives for evaluation and input, we should identify the criteria to be used in the evaluation step. For one thing, the alternatives should be evaluated against the original core ideas, values, expectations, aspirations, purposes, and requirements identified in the earlier steps.

8. Using the Criteria, Select the Most Promising Alternative

This step involves serious discussion of the alternatives and ultimately a decision about which is the most promising alternative or combination of alternative solutions. Ideally, designers and representative customers should participate in this step.

9. Describe the Future System

We now describe the future system in terms of its components. I'll say more about describing these components later in this chapter. A description of an educational system would include descriptions of the management subsystem; the instruction, curriculum, and assessment subsystems; the student management subsystem; the evaluation subsystem; as well as others. These descriptions might resemble blueprints, technical descriptions, job descriptions, manuals, and handbooks.

10. Plan for the Development of the System, Based on its Description.

In reality, we plan for developing the system at the same time as we are doing the earlier steps. Depending on the type and scope of the system we

> ## THE DESIGN PROCESS
>
> 1. Diagnose and describe the design goal.
> 2. Clarify the reason for engaging in design.
> 3. Establish the boundaries of the design.
> 4. Formulate core ideas, values, and an image of the future system that will guide the design.
> 5. Define expectations, aspirations, purposes, and requirements of the system to be designed.
> 6. Create and evaluate alternative representations of the future system.
> 7. Establish criteria by which to evaluate the alternatives.
> 8. Using the criteria, select the most promising alternative.
> 9. Describe the future system.
> 10. Plan for the development of the system, based on its description.
>
> —Adapted from Charles M. Reigeluth, Bela H. Banathy, and Jeannette R. Olson (Eds.), ***Comprehensive Systems Design: A New Educational Technology***, NATO ASI Series F Volume, ARW 900426 (December, 1990).

are designing, we might decide to develop the system ourselves using people within our own organization, or we might rely on contracts with outside specialists or groups to develop the system or some of the system components for us.

DESIGN APPROACHES AND STRATEGIES

Systems design is not an intuitive process. It is not an activity that humans naturally engage in. There are skills, approaches, and strategies for systems design that have to be learned and practiced before people become adept at this activity. Systems design requires an understanding of what design is and how it is carried out. It requires the development of design capability in the people who will be designing the new system, and it requires the organizational capacity to carry out the design activity. Much of this preparation can be accomplished through special training sessions about design approaches and methods. Some programs and materials are available to do this. (In addition to this book, other materials that can be used to develop people's design capability are listed at the end of this chapter.)

In this section, I introduce and describe two strategies and approaches for design: *idealized design* and *concurrent design*. These approaches are complementary, that is, they are used together. Elements from each approach can be combined in various ways, but they each emphasize different aspects of the design process. I also describe the use of models to help identify the components of the new system.

Idealized Design

Idealized design is thinking of how you would like the system to be if you could have it just the way you want it. Dr. Russell Ackoff, former dean of the Wharton Business School at the University of Pennsylvania, calls idealized design, "thinking of a system that does not yet exist or the idealized version of one that does."[8]

Idealized design is thinking in reverse. You begin with a vision of an ideal image of the entire system and then you determine how the organization should be structured. This ideal image should be kept constantly before you during all of the design activities. It should serve as a constant reference point for your design. Idealized design is based on the idea that one can only invent the future by working backwards from it.

Arriving at an ideal image of an entirely new system of education will involve group meetings and consensus building. This is because the image of the new system must be a shared vision created by all who will be designing the system and representatives of those who will be using it. This is to keep the design close to the expectations and desires of those who are to be served by the system. All stakeholders must be allowed to participate—teachers, parents, students, legislators, and others. Parents and students may have little time to participate in a design effort, so the design team must solicit their input and pay close attention to it because when the system is put into place, if it is not in harmony with the consumers' wants, needs, and values, it will fail.

The design teams who are responsible for designing individual components of the system will need to meet together frequently to discuss system integration, consistency, and compatibility among the components. Representatives from the design teams must also meet with all of the various stakeholder groups mentioned above to present ideas and solicit input and reactions. During this process, individuals and groups may offer recommendations, gather facts, present different opinions, and express varying viewpoints and values. Issues are resolved by arranging concerns on a scale and prioritizing them so that when a decision is reached, it represents, and takes into consideration as much as possible, the different positions of the various stakeholders. The process of arriving at a decision may involve discussion and negotiation. We do not always arrive at the solution that would appear optimal from everyone's perspective. Idealized design, therefore, becomes more complicated as the number of perspectives that must be taken into account increases.

The insert on the next page describes situations where idealized design is difficult. If the organization is faced with many of these conditions, it will be distracted easily from the design effort and the need for fundamental change. In the case where it is necessary to start in the middle of an ongoing operation (such as schools), one has to deal with the reality of the existing system. Implementation of new designs into that existing sys-

> **WHEN IDEALIZED DESIGN IS DIFFICULT**
>
> - There is a low tolerance for change.
> - The system members are not prepared to accept fundamental change.
> - There is little receptiveness to any external knowledge base or assistance from external facilitators.
> - The individuals or organization lack experience in the design process.
> - Resources to conduct the needed design process (time, energy, support, knowledge, and skills) are limited.
> - There is low tolerance for time delay (lack of organizational patience).
> - It is necessary to start in the middle of an ongoing operation.

tem must usually be done either incrementally (in gradual stages) or by establishing a new system that will function for a time in parallel with the old system.

In attempting to do redesign in education, we face many, if not all, the obstacles mentioned in the insert. We must, therefore, understand the difficulty of the task we face and realize that to be successful, we will need to involve those who are working in the current system. They will need to acquire design skills and understanding of the design process. We will also need to shift some resources to support the design effort.

The idealized design process described above may be highlighted by an example. Currently, in Florida, we are designing a system of elementary, middle, and high schools in nine collaborating school districts. This effort is known as The Florida Schoolyear 2000 Initiative. We began by forming focus groups and advisory committees at various levels: with parents, students, and teachers in the school districts that are participating with us; with administrators at the school, district, and state levels; with state policy makers, business and civic groups, and others. We prepared those working in the system to participate in a design effort by holding training sessions, talking about the design process, and forming design teams. The participating school districts and state legislature allocated resources to support the effort. As a result of this, we formulated a shared vision of an idealized system, and we set in motion the structures for doing the design work.

Concurrent Design

Because design decisions made about one part of the system affect operations in other parts, the total system must be designed concurrently—that is, simultaneously. This is necessary because systems design is a design of an entire *system* and not of isolated or independent parts. System design

Systems Design

requires that all system processes be viewed as interrelated parts, to be analyzed, designed, and developed in harmony, as an interdependent process. (Remember, interdependent means that the success of one part will depend on other parts.) To design an educational system, then, requires that the curriculum, the instructional processes, teacher training, management, and so forth, be designed at the same time. This can be accomplished by forming design teams for each of these processes. The design teams need to meet together often to present their designs and to integrate and coordinate design ideas. As the design teams deliberate and review each other's ideas, they must focus on how to make the ideal system work rather than on the reasons it won't work. You don't want to throw out a design that covers 95% of the problems because another 5% are not covered. Also, because the design will continue to evolve, all the elements proposed for the design should be treated as provisional. Many design elements are simply "placeholders"—the best solutions we have identified so far. It is expected by all that the solution will change when a better one is identified.

Models

Models are one of the key tools that we can use to help us think about a system. A model also helps us to communicate our thoughts to others. A model is simply a picture, a diagram, or a list of elements that represents, in a simplified way, the system under consideration. For example, a diagram depicting the different activities that take place in an office, factory, or school and the flow of material, documents, or people through those activities would be called a model. The model helps to simplify something that is actually quite complex. Since many elements of a system cannot be observed directly, a model helps us to visualize those elements.

Models can also be in the form of flow charts in which boxes are used to show the services or activities being rendered and arrows to indicate the routes taken by people in the system. In the simplest case, the flowchart would be linear, in the sense that work processes, products, and people flow from one stage to the next.

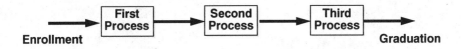

But in realistic situations, the flow may be far more complicated and might look like this:

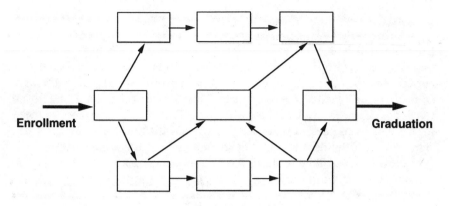

Models can also identify the parts of the system by using boxes to represent different services or activities (not in any particular order). The model below depicts possible components for an image of a system of schooling. All the services and activities described in the boxes affect student learning.

In the model, we try to include all the services or activities that are part of the system under consideration. However, this is not as easy as it sounds because subsystems are always embedded in larger systems. Sometimes it is difficult to sort out which activities are part of "our" system and which are part of the larger system. For example, the system called management operations may be responsible for budgeting or allocating funds and resources within a school. This may work very well as long as we limit our view and concern to within that particular school. But the funds are allocated to the school by a larger system that includes the school district, state, and some federal programs. These systems may not work well at all. To understand

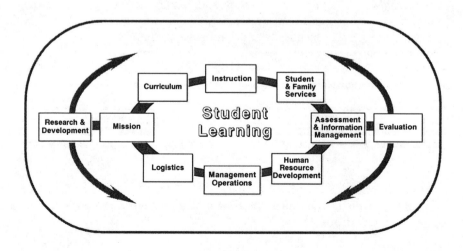

Systems Design

and explain these relationships, we have to generate an even larger model that describes the activities of the larger system. Inevitably though, as we begin to consider the larger system, the problems and complexity become enormous.

In general, the larger the system, the more parts there are to interact, the more compounded, elaborate, and intricate become the interactions, and the more difficult it becomes to understand the whole situation. For this reason, we must be careful not to get overconfident about our understanding of the system. We have to realize that our proposed design will not be perfect and may not yet be what people want or need. In the next section, I discuss the importance of creating alternatives and "counter designs" for the purpose of testing our assumptions and identifying errors in our proposed design.

COMPONENTS OF A SYSTEM

Most Americans have watched on television at least one of the NASA space shuttle launches from Cape Canaveral, Florida. To undertake and complete a space shuttle mission requires the planning, design, coordination, and effort of many people. Some may work on designing or building the physical structure of the shuttle, some on the booster rockets that will lift the shuttle out of the earth's atmosphere, some people work on the launch pad, some work on planning and budget, some are in charge of training the astronauts, some work on communications, and there are yet many other functions and capabilities that have to be performed for a successful mission. All of the actions of these different groups of people have to come together in a delicate and elegant way for a successful launch. Each ingredient of the mission must complete its individual purpose perfectly, totally, and completely if the overall mission is to be a success.

Just like a successful space shuttle mission is dependent on the delicate and intricate functioning of a number of elements and components working together as a system, education and learning also depend on a number of functions and activities. One of the most important steps in designing a system is to identify the functions and activities that have to take place somewhere in the system for it to accomplish its overall purpose. In the following discussion, I will use the word *component* to refer to the functions, activities, and ingredients that make up a system.

To identify the components of an education system, we have to think about the system in terms of various kinds of 'activities' that are performed within the system: curriculum, instruction, assessment, management, purchasing, hiring personnel, transportation, teacher training, and so forth are examples of activities that must take place in a system of education. At first estimate, the components can be represented as a list of subsystems such as the ones listed in the boxes of the previous model: mission, curriculum,

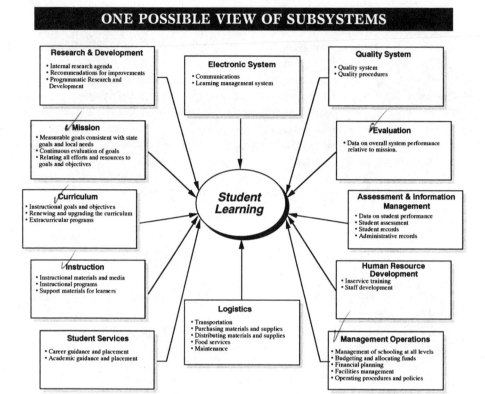

instruction, assessment, management operations, logistics, and so forth. Each of these subsystems would be responsible for carrying out various types of activities. For example, under logistics, it is easy to see that activities such as transportation, food services, maintenance, and purchasing might pertain to this subsystem. Even in this breakdown there will be some overlapping—e.g., between the activities that might come under the logistics subsystem and those that might come under the management operations subsystem. So the design group has to agree as to how to categorize each activity. This may be a fairly arbitrary decision, as long as the design group can agree and go on with the design. (See insert, "One Possible View of Subsystems.")

The system designer attempts to identify the whole system and its components, then to define alternatives in terms of how these components will function and how they will be organized. Remember, we are *designing* a system that does not yet exist. So we are free to think of new and different ways to organize the necessary activities that will go on in the new system. It is important to be creative and to generate not just one design, but vari-

Systems Design

ous alternative designs. This helps us explore more of the possibilities and think of creative ways to accomplish the purposes for which the system is intended.

The activity of creating alternative designs for the new system must involve at least a moderate amount of time and financial investment. This should not be slighted as an unnecessary activity. A large number of people and a variety of different groups should be invited to submit ideas or even specific proposals, design suggestions, or prototypes. Design competitions can be held between commercial firms and groups that represent different ideological viewpoints. Each group presents its ideas and rationale. After the competition, the best ideas can be selected for implementation. If this part of systems design is done well, the amount of time and money spent on creating alternatives can be quite high. But the payoffs (in terms of a better quality design) are also high.

The process of creating alternatives should allow, even encourage, the representation of opposing views. The process of systems design seeks to bring these opposing views into focus. Remember that the process of systems design is a scientific process and therefore should be a dispassionate one. The system designer has to give alternative ideas credibility and not get involved in the visceral reactions of the proponents. He has to sympathize with each viewpoint and examine each thoroughly, objectively, and dispassionately. This is because the system designer knows that every plan that appears to be the best is going to have weaknesses somewhere. Each alternative design will have some outstanding qualities. It is likely that most of the outstanding qualities of the various designs can be combined in the final design.

The alternative design proposals that are presented by the different people or design groups should have many of the same characteristics as the proposed design. They all must be plausible, reasonable, and attractive. They must be equally cost effective. They must incorporate all of the data that were used to build the proposed design, but may give the data different interpretations based on different views of what the whole system ought to be like. The proponents of the alternative designs must realize that the arguments for each plan are open to serious question. The purpose of the counter designs is to test the basic assumptions, to dispel unwarranted confidence, and to identify design flaws.

Ultimately, the final design will be tested in practice. If the design is for a public school or private school system, it will need to be attractive to the families that will attend there. As a rule, public schools are not used to considering themselves as competitors in the educational marketplace, but as more and more alternatives to public schools become available to students and their families, public schools will find themselves competing more and more for resources and students. The good thing about it is that this type of competition leads to a higher quality of service for all.

MISSION

All well-designed systems are directed toward a more or less specific purpose. In education, that purpose is, presumably, the education of students. But specific expectations change over time due to changes in society, in the workplace, and in the economy. The component of the system that continually defines and changes its mission is the *mission* subsystem.

The mission guides all use of time, physical resources, and personnel. Every time a dollar is spent, or a person performs a task, or a physical resource is expended, there is a lost opportunity. In other words, these resources could have been used to do some other kind of job. The objective of the mission subsystem is to ensure that every dollar spent, every task performed, and every resource expended contributes more to the real objective (mission) of the system than if it had been spent any other way. The effectiveness of the components should be measured and evaluated based on their contribution to the overall mission. The performance of the system as

---

ROCKET TO THE MOON

If we wish to land an object on the moon, then clearly we need:

- a propellant system, i.e., a mechanism capable of pushing the object out of the gravitational field of the earth;
- the design of the "ship" that will fly to the moon and the boosters that will permit it to make its flight and to land satisfactorily; and
- a communication-control system that will enable people on the earth to know where the object is and, if necessary, to control its flight, and to learn when it has landed.

If we add to the above three components a system that determines the overall objective for the system (to go to the moon), then the whole set of subsystems and their plans and their measures of performance constitute a "systems design" for the problem of putting an object on the moon. This last component, which determines the overall objectives and relates the subsystem standards to the overall, can be called the "mission" subsystem. It is the subsystem that thinks about the overall objective.

If the mission subsystem works correctly, it thinks about the relationship of the overall objective to the components at all times. It does not postpone its thinking until the system is out of sync with what is around it. It does not begin by listing a set of things it wants to do without bothering about the reason why it wants to do them. Every action within every subsystem is justified in terms of the overall objective. This does not mean that the overall system and its subsystems are rigid and closed. It means that this subsystem thinks about how it should act when the unexpected occurs.

—Adapted from C. West Churchman, ***The Systems Approach*** *(New York: Dell, 1979), pp. 8–9.*

Systems Design

DESCRIBING A SYSTEM

When we think about a system, we must think about the basic characteristics that give meaning to the system. These include:

- the purpose of the system
- the performance measure of the whole system
- the system's environment, the constraints within which the system operates (boundaries)
- the resources of the system (time, money, and people)
- the components of the system—their activities, purposes, and measures of performance
- the management of the system
- the clients of the system
- the stakeholders of the system

---

a whole should also be measured and evaluated by its success in accomplishing the stated mission.

The mission subsystem constantly compares what SHOULD happen with what DOES happen. For example, if the principal of a school says, "Our mission is to provide a quality education to each individual child and to treat the child as an individual," then our test to see whether the system is really directed toward that mission is to observe whether the system will knowingly sacrifice other goals in order to attain its stated mission. If a person or organization says that its mission is to "provide a quality education to each individual child" and yet occasionally seems quite willing to dismiss the individual needs of a child, then we have observed that the stated mission is not the *real* mission. The system has been willing to sacrifice the stated goal at some time in order to attain some other goals. To see what the system is up to requires us to observe what it actually *does,* not what it *says* it does. (We are talking here about systems that already exist. If we are designing a new system, we state the mission and then design all the functions and activities to be directed toward that mission. Once the system is operational, we can begin to observe if all activities are directed to the stated mission.)

One of the functions of the mission subsystem is to help us avoid diverting our efforts away from our mission. Suppose, for example, we notice that certain students come to school without breakfast, appear to be uncared for or neglected in their health or dental care, or lack proper clothing or materials. These are serious problems and they affect the students' ability to learn. In designing the system, we need to be aware of these problems and decide whether our mission will be to address them directly as part of our system (provide breakfast, health care, and free materials) or to consider them external to our system—the environmental conditions in which

> **EXERCISE: IDENTIFYING THE MISSION OF AN EXISTING SYSTEM**
>
> Select an organization that has a fairly well stated mission (a business, a school, a school district, a government agency, a social service agency, a hospital, etc.). Spend part of a day observing that organization to see what it does. Notice if you observe actions that go contrary to its stated mission. If you do, try to describe the REAL mission of the organization.

our system operates. In the latter case, we would devise procedures for talking with the parents about these situations and possibly referring them to other places or community agencies whose mission IS to help with these problems. Our mission subsystem reminds us that if we try to deal with problems outside of our specific mission, we will not accomplish the mission we *have* set out for ourselves.

Another function of the mission subsystem is to help us avoid applying programs to those who do not need them. For example, we might notice students who appear to need a special learning program, drug education, sex education, or who are at risk of dropping out. We might design programs, to be carried out as part of our system, to help these students, and we might be tempted to apply our programs to all students. But there might also be students who do not need this particular program or service, and to apply the program to them would be a waste of time and resources. Prevention doesn't mean directing preventative programs at all people; it means targeting those who are likely to have a problem. The idea behind prevention is to identify the potentially at risk at some early state where education and assistance could keep them from ever arriving at a latter state where the problem becomes critical.

MANAGEMENT OPERATIONS

The *management operations* component is responsible for managing the total system. It manages all the other components. Management operations must also make sure that the plans, activities, and functions are being carried out according to the system's mission. If not, it must determine why they are not. This function is called *control*. The use of the word "control" does not mean control in the sense of applying coercion to people. It means intervening when a system process gives evidence of too great a deviation from plan. The function called "control" not only examines whether plans are being carried out correctly, it also evaluates the plans, and sometimes proposes changes in plan. No person or group can set down initially the ideal design of a system or fully identify the resources or the environmental constraints and demands that will be placed on the system. Therefore, once the new system is in place, the management operations part of the system must receive information that tells it when its activities are out of sync with its mission and must include steps that provide for an orderly change back toward the mission.

RESEARCH AND DEVELOPMENT

Another important component is called *research and development*. Research simply means gaining understanding about what works and what doesn't based on objective measures of results. Development means producing or designing materials and methods that are based on our best understanding of what works.

Research and development are important because without them, money may be spent on ineffective processes and materials. Worse, the time and effort of those trying to improve learning for students can be wasted. This is because, without valid research and development, we cannot formally identify better ways to do things. We cannot know if what we are doing is producing results that are better or worse than something else.

Teachers and others working in education become cynical when they are caught up in a cycle of requirements to use untested methods that, ironically, replace older failed methods or sometimes replace something that actually worked much better. Take the example of the "open space classroom" concept that became so popular several years ago in elementary and middle schools. An ideological battle was fought between those who advocated the "open pod" style of architecture for schools as superior to self-contained classrooms. Some schools wholeheartedly adopted the open pod concept along with the team teaching approach that was part of the concept. Other schools did team teaching without the pods. Some schools with open pod architecture maintained the traditional style of teaching. Most schools eventually formed one opinion or another about the effectiveness of these changes, but in almost no cases was there any research data to support their opinions. In a very few cases, the performance of the students was measured and compared against other alternative methods. Without adequate research data, there may be changes, but they become little more than social experiments. It is simply gambling to assume that these changes will result in real improvement.

Many changes in education have not resulted in improved performance for students. A recent survey conducted by the federally funded National Center to Improve the Tools of Educators (NCITE) found that three-quarters of the teachers surveyed believed that recent school reforms had not benefitted the students the reforms were intended to help. Research allows us to scientifically test reforms or innovations and only use the ones that are shown to be truly superior.

In addition to methods, the schooling process involves books, activity guides, computer software, films, and other materials. These tools and resources influence how well or how poorly the students learn. If these materials are not of high quality, students will not learn quickly or well. It is not just a matter of examining the materials before they are purchased or adopted. They should be designed to specifications and then monitored by a series of ongoing assessments to insure that students are learning and that the materials are being properly implemented and adjusted based on student performance.

INPUTS, PROCESSES, AND OUTPUTS

Systems thinkers have found it quite useful to think of a system in terms of its *inputs, processes,* and *outputs*. Inputs are things like people (students, employees), money, and resources of various kinds. Outputs are the finished products of the system. The education system, for example, receives inputs of money and other resources. The input is transformed into buildings, teacher and administrator salaries, books, equipment, etc. The system then produces skilled and knowledgeable students as *output*. The system creates the output by enacting various processes and actions. *Processes* are things like selecting curriculum, registering students, testing and grading students, counseling students, and holding parent conferences.

Input ⟶ Processes ⟶ Output

The managers of the system (school, district, and state administrators) decide how the resources (inputs) will be used and what processes will be used to produce the output. They decide, for example, how much of the funds will be allocated to salaries and personnel, how much will be used for capital outlay (buildings, equipment, computers, etc.), and they set policies for governing the processes and procedures that are part of the system.

There is another set of conditions that are relevant to but not under the control of the managers. In part, these are considered to be constraints on the system: the limits on the resources that are available, the demands from businesses and colleges for certain characteristics and skills in the output (the graduates), and the limits on the type of teachers and other employees that are available to be hired by the school system. These environmental constraints are fixed and not under the control of the managers. They are considered to be part of the larger environment in which the system must operate. Examples of elements that may be considered part of the environ-

Systems Design

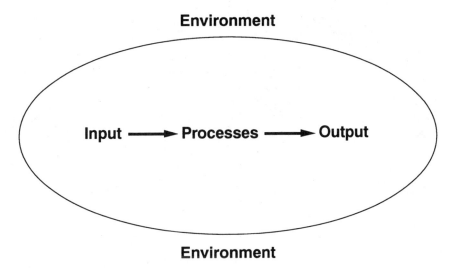

ment in which our system of schooling operates are the family, the computer revolution, the political system, and the tax structure. (See insert, "Is It in the System or in the Environment?")

How should we judge the output of the system? The output should not be judged just on the *number* of students who graduate. Our judgment has to be based on the the *quality* of the graduates in terms of the useful skills

---

IS IT IN THE SYSTEM OR IN THE ENVIRONMENT?

The environment, in systems terms, includes elements that lie outside of the system—outside of system control—but can still affect the system objectives. We can do relatively little about the characteristics or behavior of these environmental factors. The system gets requirements and demands from the environment much like an industrial firm receives sales demands from its market environment. The system also receives its raw material and new employees from its environment.

Sometimes it is difficult to tell if something should be considered part of the system itself or if it should be thought of as part of the environment. Here are some questions to help make this decision:

- Can we do anything about it?
- Does it matter relative to our mission?

If the answer to the first question is NO, and the answer to the second question is YES, then it is in the environment.

—Adapted from C. West Churchman, ***The Systems Approach*** *(New York: Dell, 1979), p. 37.*

and knowledge they have acquired and the suitability of the graduates to future employers and colleges. We also have to base our judgment on the cost of this effort.

Finally, to be a true system, we must add a feedback loop to our diagram. If the outputs of the system (the graduating students) do not meet the expected standards of quality, this must lead to some sort of adjustments or adaptations in the system processes which are directly relevant to the skill or knowledge deficiencies in the students.

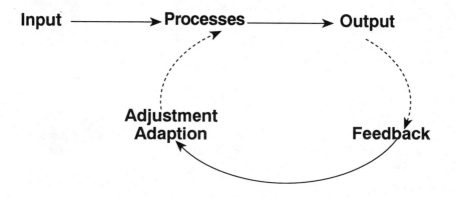

Input/output diagrams such as these are highly useful to the system designers. The total system can be diagrammed in this manner showing its relevant inputs, processes, and outputs. Each subsystem within the system also has its own set of inputs, processes, and outputs. Diagramming these (either as they now exist or as we want them to be in a new system) helps us to design processes that will lead to the desired outputs for each subsystem.

SUMMARY

Systems design allows us to create novel and innovative organizations for learning that are suited to the purpose for which they are intended. These new organizations and structures will be necessary as the boundaries of education are expanded by electronic communication, two-way video, and telecommunications. Instead of relying on the school bus or private car to take the learner to the "territory" of the teacher, we may instead rely on electronic learning networks and telecommunications to bring high quality instructional materials, human interaction, and cooperative learning directly to the student at home, at work, or in smaller, more numerous attendance centers that are open to students on flexible schedules all day, all year long. As we exit the industrial age, our educational systems, both public and private, will need to be restructured to match the requirements and characteristics of the information and communication age.

Purposeful design of these new learning organizations and structures is necessary because the current model of education that is used in most schools has reached the top of its productivity curve. Unless these organizations design new structures that individualize and optimize learning, they will remain stuck in a cycle of quick fixes, Band-aids, and piecemeal changes that result in little improvement but a lot of additional cost. The better approach involves asking the "outside view" questions discussed in this chapter and then following the design strategies outlined here to create new learning systems.

Design requires a knowledge of the design process, the ability to describe a system in systems terms, and the ability to construct and use models (component models and input/output models) as tools for thinking about our design and communicating our ideas to others. Design also requires the participation of large numbers of groups and individuals.

We can change our education systems. We can do it through systems design. Systems design is difficult and challenging, but the payoffs far exceed the effort. Together with the first technology—systems thinking—systems design has the potential to change education in the way we *want* it to change. Systems thinking and systems design, the two technologies we have described thus far, are, however, only two of the Five Technologies essential for restructuring of education and learning. Successful design of effective systems for education will require the other three technologies as well. Each of the Five Technologies is indispensable and must be used for successful school reform.

FOR FURTHER READING

Russell L. Ackoff, *Redesigning the Future* (New York: John Wiley & Sons, 1974).

Bela H. Banathy. Instructional Systems Design. In R. M. Gagné (Ed.), *Instructional Technology: Foundations* (Hillsdale, NJ: Lawrence Erlbaum, 1987).

Bela H. Banathy, *Systems Design of Education: A Journey to Create the Future* (Englewood Cliffs, NJ: Educational Technology Publications, 1991). See especially pp. 19–23 about models and their usefulness.

Bela H. Banathy, *A Systems View of Education: Concepts and Principles for Effective Practice* (Englewood Cliffs, NJ: Educational Technology Publications, 1992).

I. Cross, *Developments in Design Methodology* (New York: John Wiley & Sons, 1984).

Robert L. Flood and Michael C. Jackson, *Creative Problem Solving: Total Systems Intervention* (New York: John Wiley & Sons, 1991).

J. C. Jones, *Design Methods: Seeds of Human Future* (New York: John Wiley & Sons, 1980).

George J. Klir, *Architecture of Systems Problem Solving* (New York: Plenum, 1985).

G. Nadler, *The Planning and Design Approach* (New York: John Wiley & Sons, 1981).

C. Nigel, *Developments in Design Methodology* (New York: John Wiley & Sons, 1981).

Charles M. Reigeluth and Robert J. Garfinkle (Eds.), *Systemic Change in Education* (Englewood Cliffs, NJ: Educational Technology Publications, 1994).

Peter M. Senge and Colleen Lannon-Kim, "Recapturing the Spirit of Learning Through the Systems Approach," *The School Administrator* (November, 1991), pp. 8–13.

NOTES

1. George Leonard, "The Great School Reform Hoax: What's Really Needed to Improve Public Education?" *Esquire* (April, 1984), p. 50.
2. The following sources provide evidence for the "S-curve" phenomenon in organizations: R. Foster, *Innovation: The Attacker's Advantage* (London: Macmillan, 1986); J. R. Womack, D. T. Jones, and D. Ross, *The Machine that Changed the World* (Cambridge: MIT Press, 1990); T. Kuhn, *The Structure of Scientific Revolutions* (Chicago: University of Chicago Press, 1970); A. Levy, *Transforming Organizations: Second Order Planned Change* (University of California, Santa Barbara: Dissertation, 1985).
3. Harrison Owen, *Leadership Is* (Potomac, MD: Abbot Publishing, 1990), p. 114.
4. The necessity of making S^1 to S^2-curve transitions is described in P. W. Braden, M. J. DeWeaver, and L. Gillespie, "Keeping U.S. Companies Competitive," *Performance Improvement Quarterly*, 4, No. 4 (1991), pp. 47–61. Also in William H. Davidow and Michael S. Malone, *The Virtual Corporation: Structuring and Revitalizing the Corporation for the 21st Century* (New York: HarperCollins, 1992). See especially Chapter 2.
5. For evidence that education is stuck on an S^1 curve, see Robert K. Branson, "Why the Schools Can't Improve: The Upper Limit Hypothesis," *Journal of Instructional Development*, 10, No. 4 (1988), pp. 15–26; Larry Cuban, "The Fundamental Puzzle of School Reform," *Phi Delta Kappan*, 69, No. 5 (1988), pp. 341–344; and, Larry Cuban, "Reforming Again, Again, and Again," *Educational Researcher*, 19, No. 1 (1990), pp. 3–13.
6. These "inside-view" and "outside-view" questions were formulated by Charles M. Reigeluth, Bela H. Banathy, & Jeannette R. Olson in *Comprehensive Systems Design: A New Educational Technology*. NATO ASI Series F Volume, ARW 900426 (December, 1990), pp. 6–7.
7. Charles M. Reigeluth, Bela H. Banathy, & Jeannette R. Olson (Eds.), *Comprehensive Systems Design: A New Educational Technology*.
8. Russell L. Ackoff, *Creating the Corporate Future: Plan or Be Planned for* (New York: John Wiley & Sons, 1981), p. 105.

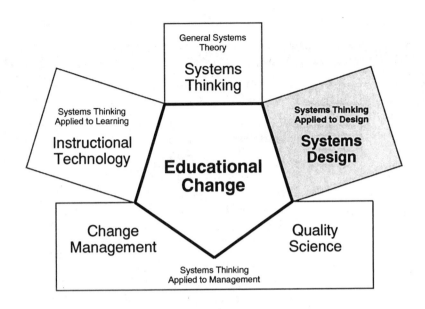

CHAPTER 5

SYSTEMS DESIGN TOOLS

I have already described models, flowcharts, and input-output diagrams which, in and of themselves, are tools for systems design. These tools help us think about the various functions and components of the systems we are designing, and they allow us to effectively communicate our ideas to others in systems terms. In this chapter, I present some other useful tools for systems design. The first is a series of design exercises that can help one develop skills in systems design. You can perform these exercises as individuals or in groups.

Once you have completed these exercises and feel that you can think systemically about these types of problems, you can try thinking about how you would go about designing an educational system. Here are some of the questions you might ask yourself:

How does the new education system incorporate:

- rewards and incentives?
- collaborative learning?
- instructional games?
- computer simulations?
- distance education?
- peer tutoring?

EXERCISES IN SYSTEMS DESIGN

To develop your skills in systems thinking and systems design, complete the following exercises:

Describing Systems Describe the following systems by identifying the components, resources, inputs, processes, and outputs:

- urban rapid-transit system
- school system
- dry cleaning laundry system
- grocery-store system
- your own household

Now develop a model that depicts the elements you have described.

Automobile Imagine that, because of some quirk in man's inventive capabilities, no faster or more convenient mode of transportation had occurred up to now than the horse and carriage, though all other technologies were up to date, including communication. Now a bright young engineer invents the combustion engine. Granted an excellent ability to forecast consequences of the type we know so well today (accidents, speeding, smog, convenience), what is a good approach to the "new" transportation system?

Privacy Given that we are now living in an information age, design an approach to information distribution and access that would value privacy. "Privacy" in this exercise means dignity and respect, independent of income and status. Complete privacy also means that the individual is not investigated and information about him is not recorded, even though he himself may not object.

Design an Information System Given that the computers of 20 years from now will be smaller, cheaper, extremely fast, wireless, and have voice recognition and voice output, describe the implications for library systems, newspapers, advertising, home schooling.

—These exercises are adapted from C. West Churchman, **The Systems Approach** (New York: Dell, 1979), p. 235.

You would also want to answer the following questions:

How does the new system allocate tasks for students, staff, and support personnel?
How are staff hired, trained, and developed?
What are the different roles of the staff members?
How is the instruction delivered?
Is instruction delivered remotely (homes, libraries, shopping malls) or only within the school site?
How long is the learning day? Year? Why?
What does the physical configuration look like?
Where are the people and what are they doing? Why?
How will success of the system be evaluated?

FUTURES WORKSHOP

The second tool is a group-involvement technique that consists of a series of prescribed activities leading to a design and action plan. The technique is known as the futures workshop.[1] It is useful for shedding light on a problematic situation, generating visions about the future, and discussing how these visions can be realized. The futures workshop starts with a critique phase, which consists of brainstorming on current problems within the current system and criticism of current practices. The participants' critiques are formulated as short statements and written as a few key words on wall charts. Speaking time is restricted to 30 seconds to make it easier for all participants to speak. The participants do not need to defend or offer arguments to support their criticisms. This enables less verbal participants to jump into the process. The insert "Futures Workshop: Criticism Phase" gives an example of the type of criticisms that might arise from a group of people designing an education system during this phase of a futures workshop.

FUTURES WORKSHOP: CRITICISM PHASE

The school is run like a factory
Too many kids don't learn
Too many interruptions
Not enough textbooks
Lack of computer equipment
We don't have a well planned curriculum
Lack of financial rewards for staff success
Too many state and federal policies

About midway through the critique session, statements are grouped under categories such as curriculum, school environment, incentives, and materials. Then, the group divides into smaller groups of four or five people each. Each group chooses a different category of short statements, and through discussion, reformulates these into a concise critique of the current situation. Then, in the final session, the total group meets together to present and discuss the concise critique statements of the smaller groups.

The second phase of the futures workshop is the fantasy phase. The objective of this phase is to invert the critical statements from the critique phase into positive statements. Another activity of this phase is to allow groups to draw pictures of the new system as they would like to see it in the future. The drawings are hung on the wall and a brainstorming session ensues. Again, long speeches are prohibited and, following the rules of brainstorming, no evaluative statements are allowed (such as, "That is completely unrealistic"). "Weird is wonderful" is the theme of the brainstorming session. No statements about future situations and future possibilities are considered too weird or extreme. Again, the statements of the participants are recorded on wall charts by the facilitator. Some examples of such statements are shown in the insert "Futures Workshop: Fantasy Phase."

FUTURES WORKSHOP: FANTASY PHASE

Telephones and modern telecommunications equipment are available
Differentiated staffing
Individual progression based on performance
Plenty of computer workstations for students
Flexible scheduling
Disciplined environment
Faster students can progress at a faster pace
Slower students get extra help
The finest educational materials and programs are available

During the final portion of the fantasy phase, the statements are ranked by giving each participant five votes to cast on the statements they most favor. The statements getting the highest scores are included in the group's "vision of the ideal system."

The third phase of the futures workshop is the implementation phase. This phase is started by first dividing the participants into groups and having each group discuss and formulate its version of the ideal system, based on the "vision of the ideal system" of the fantasy phase. After a period of time, each group presents its description to the full group for evaluation.

Systems Design Tools

The full group examines each of the descriptions to determine whether it is possible to realize it under current conditions. In addition, the group discusses whether it is necessary or possible to establish new conditions under which the ideal system *could* be realized. Suggestions as to how the visions could be realized are welcomed and encouraged. The implementation phase concludes after the descriptions are turned into a unified strategy for design, with the participants making detailed plans for how the first steps in the strategy should be carried out within a given time period, and listing the names of the people who are to perform these activities.

Futures workshops form a solid basis for systems design. Plans for specific actions are an important outcome of the workshops, and the wall charts provide crucial documentation of the ideas discussed.

BACKWARD MAPPING

A third design technique is backward mapping. Backward mapping refers to thinking about the ideal and working back from that point to determine how to create that ideal. This is similar to the term "idealized design," which was discussed in the previous chapter. Creating an ideal design for a school or educational system is a unique and creative act. Generally, a group of people work together to create the design. What you create initially does not need to be a complete or highly refined version. It is much like the preliminary sketches made by an architect. This will become your blueprint for the ideal design. Backward mapping then helps you think of the functions that will need to take place to support the design at the administrative level, school level, classroom level, and at the individual learner level. This then becomes an organizational arrangement of the people, processes, and sub-organizational units that support the idealized design.

After formulating a description of an ideal design (see the insert "Toward the Ideal School"), the next step is the backward mapping process. The purpose of the backward mapping process is to help you identify the functions that will need to be designed at the other levels of the system to support the ideal design. Functions that occur at the administrative level will influence functions at the classroom level, and so on. For each level, the design team must answer the question: "What functions must be performed at this level?" Using the insert "Backward Mapping Template," the design team tries to fill in each of the cells in the template with the functions that must be performed at that level.

Once the design team has identified the critical functions to be performed at each of these levels, it can proceed to think about jobs or groups within the current management structure that have a close approximation to one of these functions. These groups can then be oriented toward performing the new functions. In addition, some new management structures

TOWARD THE THE IDEAL SCHOOL

Before doing backward mapping, you first have to create an initial design. Here are some suggested procedures for creating an initial design:

- Never be critical of the first ideas you or your colleagues develop. Keep building on them and try to elaborate and refine them.

- Look for analogies as a way to stimulate your thinking, e.g., "an excellent school should be like a . . ."

- Don't be too logical. Just because it doesn't seem like an innovative idea, don't give up on it. By the same token, don't force yourself to be too practical.

- Ask yourself lots of "what if" questions. What if money wasn't a problem? If you could start your own private school or charter school, what would it be like?

- Leave things a bit messy at first. No effective design will be without paradoxes and ambiguity. (The current design certainly isn't without them.)

- Don't hesitate to go beyond your area of expertise. Change frequently comes from someone who's not wedded to the status quo. For example, ask some business management experts how they would organize a school.

- Avoid "group think." Deliberately challenge logic that everyone seems to agree with.

- Give yourself some rest periods in between design work sessions. Free your mind.

- Assign group members to read about someone else's effort to create a new design or "blueprint" for an innovative school.

- Consider using questionnaires and survey instruments to solicit input to the design from "customers" (students, parents, business community).

—Adapted from C. L. Hutchins, *Achieving Excellence: An Educational Decision-Making and Management System for Leadership, Efficiency, Effectiveness, Excellence* (Aurora, CO: Mid-Continent Regional Educational Laboratory, 1991).

Systems Design Tools 83

BACKWARD MAPPING TEMPLATE

	State Level	District Level	School Level	Classroom Level	Student Level
Mission					
Curriculum					
Instruction					
Assessment					
Administration					
Student Services					
Other....					

and some new job roles may need to be created to perform functions that are not present in the existing system.

CONSENSUS BUILDING TOOLS

An important part of design is building consensus among the participants on what the ideal design ought to look like. Again, it is possible, even desirable, for the design process to result in competing designs. Ultimately, however, the best elements of the various designs will need to be brought together in a final design. As design groups move toward consensus, it is helpful to keep several points in mind.

First, consensus means general agreement. For consensus to exist, it is not necessary for each group member to agree fully. But it *is* necessary for every member to be heard and, in the end, for none to believe that the decision violates his or her strongest convictions. It is not necessary that every member consider the decision to be absolutely the best one, but each member should be able to accept it on the basis of logic and feasibility. As long as everyone is willing to accept the decision (even though each member may not agree with it entirely) and to support it through appropriate actions, consensus has been reached and the judgment can be considered a group decision.

One helpful tool in consensus building is the consensus grid. The design team lists potential problems or issues along the left side of the grid. Then each person uses his or her own copy to rank the issues in terms of

CONSENSUS GRID

Statements	Sums of Individuals' Ranking 1 2 3 4 5 6 7 8 9 10	Consensus Ranking

their perceived importance. The sums of the individual rankings are computed on the consensus grid to obtain the consensus rank. The group may choose the top two, three, or four for further discussion and re-ranking, or decide to accept the ranking as is. This technique shows the group where consensus does or does not exist and moves the people along more quickly toward a consensus decision.

Another helpful tool for building consensus is a simple form that can be used to solicit input and reactions to a design decision. Since people can agree with a decision and still have some reservations or concerns about it, several levels of consensus are specified on the form: Level 1—"I agree with the decision and will actively support it"; Level 2—"I can live with the decision"; and, Level 3—"I do not agree with the decision and cannot support it." Note that both Levels 1 and 2 represent consensus. The form also allows individuals a way to indicate whether or not they feel that they had sufficient opportunity to provide input or to request additional information. On another area of the form, they can indicate any concerns that they still have about the decision. The forms can be used to solicit input anonymously, or individuals may choose to identify themselves. The form is a useful way for the leader to assess the true feelings of participants in a design effort.

GROUPWARE AND COMPUTER DESIGN TOOLS

Groupware is the name used to describe computer software that links people together and allows them to work on a common task. Recently, a number of software companies have released software tools designed to

CONSENSUS PROCESS FORM

Action: _____
_____
_____
_____
_____

Date Initiated: _____

Name: _____ Group: _____

- ❑ I have had an opportunity to provide input, and to present problems or issues related to this action.

- ❑ To arrive at consensus on this action, I would need the following additional information:

_____
_____
_____
_____
_____
_____

- ❑ I have the following concerns about this action:

_____
_____
_____
_____
_____

Consensus Level:

- ❑ Level 1: I agree with the decision and will actively support it.
- ❑ Level 2: I can live with the decision.
- ❑ Level 3: I do not agree with the decision and cannot support it.

help teams do design work and group decision making. These "electronic meeting" tools can be used very effectively to help teams do design work.

Most of these electronic meeting tools use a combination of computer software, individual handheld remote numeric keypads, and a projection system. The team members use their keypads to respond to questions posed by the team leader or facilitator. Their answers are automatically processed by the computer and compiled in easily interpreted graphs on a large screen. Because it allows everyone to get involved and helps keep the group focused, this type of groupware is a remarkably effective way to stimulate discussion, even in meetings populated by compulsive talkers and habitual listeners. Everyone has an equal voice with the keypads and so the quiet participants feel more empowered and the talkers are more quiet. Team members can vote on alternatives with the push of a numbered button. They can rank alternatives in order of priority. They can indicate how they feel about a particular proposal or decision on any scale they wish to construct. For example, people might be asked to respond to the question: "How comfortable are you with this decision?" and then respond on a five-point scale ranging from "very comfortable" to "very uncomfortable." They can also use agree/disagree scales to determine agreement or consensus on design decisions or proposals.

OPTIONFINDER KEYPAD

—Photo courtesy of Option Technologies, Inc.

Systems Design Tools

Most important decisions in systems design work are made in meetings. Yet we have all experienced meetings in which there are significant barriers to open communication and group decision making. For example:

- Input from all participants is time consuming; the larger the group, the greater the problem.
- A few participants tend to dominate.
- There is no opportunity for private communication of concerns.
- It is difficult to record participants' views.
- It may be difficult to maintain the attention of non-active participants.
- Participants with limited communication or group process skills may exclude themselves or be excluded.
- It is difficult to assess group consensus.

One of the main values to using electronic group software is that this tool helps to overcome many of these barriers by providing meeting participants with equal opportunity for input. It also collects, organizes, displays, and records participant responses for viewing.[2]

Another computer tool that is useful in the design process is *Inspiration*—a computer software package based on the concepts of mind mapping, clustering, and idea mapping. Inspiration helps you brainstorm your thoughts, diagram information, and create design maps. Inspiration has both a diagram view and an outline view. You use the diagram view for visual brainstorming and diagramming, and the outline view for organization and writing. *Inspiration* integrates these two views so you can switch from one to the other effortlessly.[3]

In an *Inspiration* diagram, you write your ideas, concepts, or information within a *symbol*. To show relationships or order, you connect symbols with *links*. Each symbol has a text area, which appears inside a dotted rectangle within the symbol. Each symbol has *handles* so that you can "grab" it to add links or change its size. See the insert on the next page.

COMPUTER SIMULATIONS

At this writing, I know of no computer simulations that can easily be used to simulate a school or educational system. However, there are several excellent computer simulations that can be used to simulate other human activity systems. These computer simulations can be used by individuals or groups to practice the art of design as applied to other problems. These engaging simulations are a great way to develop initial concepts about design. The learning that results from use of these simulations can then be discussed and applied to designing an education system.

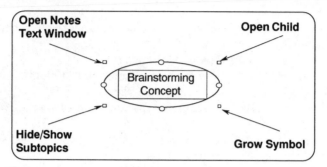

- Rectangular-shaped handles to draw links.
- A handle at the upper right to open a *child* (sublevel diagram or outline) for the symbol. A handle highlighted in gray indicates there is a child.
- A handle at the lower right corner to increase or decrease the size of the symbol.
- A handle at the upper left corner to open the text window for a symbol. The Notes Text window is used to write about your idea. A handle highlighted in gray indicates that Notes Text exists for the symbol.
—From ***Inspiration 4.0 User's Manual***, 1st ed. (Portland, OR: Inspiration Software, Inc., 1992).

SimCity can be used to design and build a model city. To use *SimCity*, you start with an empty topographical map and then proceed to make a variety of decisions resulting in the development of a city that grows or dies based on the quality of decisions made. First, you are required to decide where to put your city. To make this decision, you need to take into consideration a variety of factors and weigh them to determine an appropriate location. Then you are asked to determine the ideal population size, to select the appropriate source of electrical power, to determine the best transportation systems, to adopt or not to adopt zoning regulations, to determine the number and types of schools and hospitals required, to decide what types of industries you'd like to attract and how to do that, and to consider many other factors.

As the group uses the simulation, the group needs to ask and answer thousands of questions to complete the design of the city. How many people should there be in an ideal city? How can we find out? What cities appear to be the best designed today? How big are they? Where are they located? Should we use coal, nuclear, or hydroelectric power plants? How many schools do we need? What percent of the population is school age? What percent will be in elementary schools? High schools? Should we have a college? A movie theater? Athletic facilities? Where will people work? How will they get there? Should all of the industry be placed together or should it be spread around? How much trash will be produced? What will we do with it? How many fire stations, police stations, hospitals? Which activities and services will be funded through taxes and which by the users or consumers? And many, many more. Finally, the users build a scale model of their city based on their answers to these questions.[4]

SUMMARY

The tools and techniques for systems design illustrated in this chapter are basic to the process of systems design. Systems design exercises and futures workshops are very useful in getting a group of people oriented to the process of design and in showing how *design* is different from simply changing or adapting something that already exists. Backward mapping and use of the consensus grid and consensus process form are ways to move a design team through the initial brainstorming process to the point of having a viable design or blueprint for a new system. Groupware and computer design tools like *Inspiration* provide a modern, high-tech way of accomplishing some of the design process. Computer simulations like *SimCity* can be used for practice in thinking about the process of design.

FOR FURTHER READING

Bela H. Banathy, New Horizons for Thorough Systems Design, *Educational Horizons,* 69, No. 2 (1991), pp. 83–89.

Bela H. Banathy, *Designing Educational Systems: Creating Our Future in a Changing World.* In Charles M. Reigeluth and Robert J. Garfinkle (Eds.), *Systemic Change in Education* (Englewood Cliffs, NJ: Educational Technology Publications, 1994).

C. West Churchman, *The Systems Approach* (New York: Dell, 1979).

Diane M. Gayeski, *DesignStation 2000:* Imagining Future Realities in Learning Systems Design, *Educational Technology,* 35, No. 3 (May–June 1995), pp. 43–47.

C. Larry Hutchins (Ed.), *Achieving Excellence* (Aurora, CO: Mid-Continent Regional Educational Laboratory, 1992).

Daniel H. Kim, *Systems Archetypes I: Diagnosing Systemic Issues and Designing High-Leverage Interventions, Toolbox Reprint Series* (Cambridge, MA: Pegasus Communications, 1992).

Daniel H. Kim, *Systems Archetypes II: Using Systems Archetypes to Take Effective Action* (Cambridge, MA: Pegasus Communications, 1994).

Carl M. Moore, *Group Techniques for Idea Building,* 2nd ed. (Thousand Oaks, CA: Sage, 1987).

Roy Payne and Cary L. Cooper (Eds.), *Groups at Work* (New York: John Wiley & Sons, 1981).

NOTES

1. Futures Workshops were originally developed by Robert Jungk and Nerbert Müllert for citizen groups with limited resources who wanted a say in the decision-making processes of public planning authorities (town planning, growth management, roads, energy crisis, etc.) See Robert Jungk and Nerbert Müllert, *Future Workshops: How to Create Desirable Futures* (London: Institute for Social Inventions, 1987). Finn Kensing has written about use of this technique in system design. See Finn Kensing, Generation of Visions in Systems Development, in P. Docherty, K. Fuchs-Kittowski, P. Kolm, and L. Mathiasen (Eds.), *Systems Design for Human Productivity—Participation and Beyond* (Amsterdam: North-Holland, 1987), pp. 285–301. Another key resource on Futures Workshop is Marvin R. Weisbord, *Discovering Common Ground: How Future Search Conferences Bring People Together to Achieve Breakthrough Innovation, Empowerment, Shared Vision, and Collaborative Action* (San Francisco: Berrett-Koehler, 1992).
2. For anectodal evidence of the value of electronic group software, see W. M. Bulkeley, "'Computerizing' Dull Meeting Touted as an Antidote to the Mouth That Bored," *The Wall Street Journal*, January 28, 1992, B-1; A. R. Dennis, J. F. George, J. M. Jessup, J. F. Nunamaker, and D. R. Vogel, "Information Technology to Support Electronic Meetings," *MIS Quarterly*, 12, No. 4 (1988), pp. 591–619; S. L. Jarvenpaa, V. S. Rao, and G. P. Huber, "Computer Support for Meetings of Groups Working on Unstructured Problems: A Field Experiment," *MIS Quarterly*, 12, No. 4, pp. 645–665.
3. *Inspiration* is published by Inspiration Software Inc., P.O. Box 1629, Portland, OR 97207.
4. *SimCity* is published by Maxis, Inc., 2121 N. California Blvd., Walnut Creek, CA 94596.

ELECTRONIC MEETING AND DECISION MAKING TOOLS

Companies that are currently marketing electronic systems for meetings and consensus facilitation include: Option Technologies, Inc., 1275 Knollwood Lane, Mendota Heights, MN 55118, tel. 612-450-1700; Wilson Learning Corporation, 7500 Flying Cloud Drive, Eden Prairie, MN 55344, tel. 612-928-2880; ComTec, Inc., 180 Summit Ave., Montvale, NJ 07645, tel. 201-476-0200; and Audience Response Systems, Inc., 2148 N. Cullen Ave., Evansville, IN 47715, tel. 812-479-7507.

PART III
The Third Technology: Quality Science

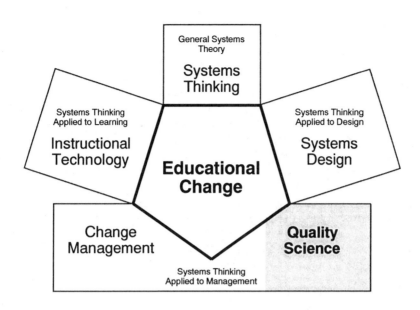

CHAPTER 6

QUALITY SCIENCE

"If there is a better way to do it . . . find it."
—**Thomas Edison**

Quality science is the application of systems thinking to management and to the problem of producing a product or service that satisfies the customer's need each time, every time. Some readers who are already familiar with quality science may not have thought about it in this way—as an application of systems thinking. But in reality that is exactly what it is. The roots of quality science go back to early operations research and statistical quality control methods used in manufacturing during World War II. These methods arose at about the same time and in combination with systems thinking. Operations research was an early application of systems thinking concepts to manufacturing. Since those early days, these methods have expanded to include the modern quality methods of today. Quality science is based on a body of knowledge that includes an understanding of systems theory. To understand quality science, one must first understand that organizations work as systems. Their parts are interdependent and synergistic.[1]

Today, most businesses, large and small, have adopted some of the quality science approaches to the conduct of their business. Many branches of government at both the federal and state levels have introduced "Total Quality Management" (TQM) into their day-to-day operations. In 1990, the federal government initiated the Malcolm Baldrige National Quality Awards to recognize "world class" accomplishments in the area of quality. Some state governments have also initiated award programs (such as the Sterling Award in Florida) to similarly recognize high levels of accomplishment in applying quality science.

Quality science is without question the single most important management development in business and manufacturing to come out of the twentieth century. It is the single most important thing we can point to that has allowed American businesses to compete with their global competitors and regain the market share that they had previously lost. Quality science has been remarkably successful in other settings where it has been seriously put to use, and it can be just as successful in education.

BASIC CONCEPTS

The basic idea behind quality science is that quality control can be designed into every procedure, every action, every step of every process that occurs in an organization. Instead of inspecting for quality at the end, we can continually improve the actions and steps *throughout the process*, thus assuring a much better result each time. We must always be striving to improve each and every process.

The idea that work can be broken down into processes is fundamental to quality science. These processes involve an integration of people, material, methods, and machines. The process of manufacturing a product consists of a variety of sub-processes, procedures, and actions. For example, the work of providing medical care in a hospital likewise can be described as different processes, procedures, and actions. The same is true for the work done by a travel agency, employment service, government agency, or school. By studying the steps, the materials, the machines, and the actions of the people involved in the processes, we can continuously improve those processes.

If a system is not producing the results that we prefer, it is because the processes we are using aren't working well or we have chosen the wrong processes. Let me give an example. Suppose we were managing a manufacturing company and a large percentage of the products that result from our production process have to be scrapped, reworked, or warehoused, all at additional cost. In this case, it would be ludicrous to suggest that the solution to our problem is to run the same production process more hours each day, or to pay the workers more, or to have higher inspection standards at the end. Instead, we should carefully examine all of the processes involved in

> **EXAMPLES OF PROCESSES**
> - Registering for classes
> - Preparing a science project for the district science fair
> - Processing a purchase order
> - Getting ready to go to school in the morning
> - Solving a problem
> - Writing a term paper
> - Teaching fractions
> - Handling parent questions or complaints
> - Monitoring attendance
> - Holding a parent/teacher conference
> - Dealing with discipline problems
> - Scheduling an appointment to see a counselor
> - Advising students on applying to college

production, scrutinize all the different stages of production, identify which processes are not working well, and look for better processes.

The quality science approach asserts that work is not haphazard. It can be studied, analyzed, and scientifically dissected. It requires monitoring and measuring all of the processes inside the organization to see if they are adding value and meeting customer needs. These observations and measures are used to guide decisions about how to improve the actions, the processes, and the procedures that are used by people in the organization. Thus, the intent is not to control or blame people for problems in the organization. Rather, it is to identify what *works* and assure that every action and procedure used is the best and most effective way currently available.

Quality science is also the key to controlling costs. Costs can be reduced by substituting more efficient processes in place of less efficient ones. We have to stop spending money on steps, actions, or processes that are less effective and start spending that money instead on actions that produce better results. This is why quality science is absolutely essential for success in educational restructuring. It is one of the Five Technologies because the others cannot be implemented without it. Using a quality science approach is what makes the whole thing affordable! Without it, we continue to spend our dollars on less than effective processes. We cannot expect improvement and at the same time continue to expend budget dollars in the same ways as in the past. We have to know what works and what doesn't and then direct resources to what works and stop spending money on what is known not to produce good results. Quality science is the technology that provides reliable answers to the questions: What works? Where should we spend our money?

IN THE BEGINNING, THERE WAS—VARIATION

To understand quality science, we first have to understand *variation*. Variation refers to the fact that there are and will always be differences among things. In education, there is variation among schools, textbooks, equipment, and physical facilities, in room temperature, in noise level and distractions, and in other circumstances. People, of course, differ in a variety of ways. They differ in their interests, in the ways they learn best, in their ability to grasp abstractions, and in their speed of learning. They differ in preferences, opportunities, motivation, and work habits. There is variation in both students' and teachers' performance. Variation will exist in all people and environments. This idea, as simple as it sounds, is a critical concept in quality science.

The goal of an organization that is striving to improve the quality of its products and services is to identify the *causes* of variation within the system so that they can act on them. Knowing the sources of the variation allows managers to react appropriately to the performances and results they deal with daily. In a school, understanding the sources of variation in student or staff performance will allow school leaders to take appropriate actions rather than inappropriate ones. Why is it so important to understand the sources of variation? Because variation comes from two very different sources, and the actions that should be taken in response to each kind of variation are very different.

Some variation exists simply because things differ naturally. Students learn at different rates, have different levels of motivation, intelligence, backgrounds, and physical abilities. Some students may have a stomach ache or are over-tired on a particular day. No two students, teachers, or principals are identical. This is called *random* variation. It will always exist. It is not a problem in any sense of the word. With no variation, the world would be monotonous and uninteresting.

It is the other type of variation that is so important. This type of variation is called *special cause* variation. Special cause variation is variation that exists because the system *causes* it to exist. This type of variation is caused by something that the system does and the processes it uses. A process for manufacturing car tires, for example, may produce some tires that are substandard. In this case, the substandard tires are *caused* by certain flaws in the production process. Some students may not learn algebra as well as others. This may be due to random variation (different levels of motivation, different backgrounds, etc.). But it may also be due to the instructional processes or steps that were used. Perhaps the instruction did not present some necessary information. Or perhaps it was presented in a way that was not understandable to some students. Or perhaps not enough practice time was allowed. In these cases, we can track down the problems in the instruction and change them so that next time, a larger percentage of the students will succeed.

Special cause variation is "special" because it has a particular, identifiable cause. It can be tracked down and attributed to a particular action or set of actions within the system. Random variation, on the other hand, is difficult to track down and can't ever be entirely assigned to any particular source or cause.

Distinguishing between random variation and special cause variation is critical for the manager because appropriate action is quite different for each case. Without this distinction, managers may make unnecessary and costly adjustments in an attempt to compensate for random variation, thus diverting people's attention away from more pressing problems that *could* be effectively addressed. Only when managers deliberately focus on special cause variation can they take actions that will result in real and lasting improvement. Random variation can never be eliminated, therefore, managers who attempt to act on random variation run up and down unproductive paths. They also run the risk of lowering morale in the organization because people don't see any improvements from their actions, and they consequently lose confidence in their managers.

RANDOM AND SPECIAL CAUSE VARIATION IN SCHOOLS

When a lot of variation in a system is due to special causes, the system is said to be unstable, since we cannot predict what the system will produce. Its capability will be sporadic and unpredictable. This, in large part, is exactly the situation within schools today. Although some of the variation in student performance is due to normal differences among students, much of it cannot be blamed on the students. *The system processes themselves are at fault and need to be improved and corrected.*

Take the problem of students dropping out of school. To some degree, the drop-out problem is due to the natural differences that exist among students. Some students are less motivated or less interested, some are less capable, some drop out to go to work due to extreme financial challenges within their families. These conditions will continue to exist for some students no matter what schools do. But much of the drop-out problem *can* be blamed on schools. Students may have been directed into inappropriate courses; some may have been exposed too often to instruction that was boring, incomplete, or inappropriate for them. Applying quality science means that we seek to identify the processes within the system that contribute to the problem and then to systematically improve these processes.

Current managerial strategy in most schools focuses much too much blame on students or teachers. Rarely are system processes examined to identify how they can be fixed. Principals are often taught to reward teachers who perform "above average," and punish those who rank lowest. People may even be blamed for random variation over which they have no control, with the result that their jobs or careers may be put in jeopardy. What this shows is a lamentable lack of understanding of random and special cause variation.

What these principals overlook is that an average is simply a number calculated from data. Because of how it is calculated, roughly half the people in any group will perform below average, no matter how smart or talented they are. The people in the below average group may have done nothing different than those in the above average group. They were merely working within a system that contains random variation. For them to be unfairly labeled "inferior" has a demoralizing influence on the entire staff.

Think of an Olympic athlete. It is likely that a particular athlete, no matter how good he or she may be, will finish some events in the bottom half. Does this mean that we should then label this person as an inferior athlete? Should we direct additional resources toward better training of this particular person? If we did that, we would be directing resources and blame at a non-problem.

But, you say, some teachers really are better than others. Of course they are! The point is what to do about it. The manager who understands random and special cause variation will look carefully at the teachers who perform consistently below the levels of acceptability and attempt to identify any special causes that may be contributing to their low performance. The manager will investigate how one person's case differs from those of others working in the same situation. Perhaps this person was not as effectively trained or has a physical limitation (height, vision, hearing) that influences his or her ability to perform particular aspects of the job. If some specific problems such as these are identified, then the manager and teacher can work to eliminate, if possible, the specific cause.

The manager can also work to bring *everyone's* performance up to a higher level by giving better tools and training to all. If all the teachers are within the acceptable range, working on those who are below average simply because of random variation is not effective and has a profoundly negative psychological effect on those individuals and on the organization as a whole.

How do we identify whether we have random variation or special cause variation? The strategy is actually quite simple. It is done by collecting timely data and by drawing certain kinds of charts and diagrams. These special charts and diagrams allow us to tell the difference between random and special cause variation. I describe these charts and diagrams in the next chapter.

OBJECTIONS TO USING QUALITY SCIENCE IN SCHOOLS

Unfortunately, many educators and education leaders regard any idea or concept coming out of the industrial/business sector as foreign. They feel that we can't apply methods from the corporate world because "they produce goods like automobiles—and schools deal with human beings." Actually, this is a hollow argument because even industries that produce goods like cars and telephones also deal with educating and developing human beings—their employees. They have extensive training programs, personnel incentive efforts, and personal development workshops. In addition, many service sector businesses (health care, financial services, legal services, libraries, travel, and tourism) are really very similar to education in that they deal day-to-day with people. Their mission is to help people accomplish their personal goals.

Some educators don't like the words that are associated with quality science; words such as *customer, supplier, product,* and *workers.* They object, for example, to the idea of thinking of learners as "workers." They are not like workers, they say, "because workers are paid and students are not." But volunteers in a hospital are workers, the boys or girls in a scout troop are the workers in those organizations, and the members of a little league baseball team are workers even though none is paid. Although the wording associated with quality science approaches has a strong business and manufacturing orientation, it can be easily adapted to service organizations such as schools, government, and other activities.

Many of the same elements that characterize medium or large corporations are present in any school district: million or multimillion dollar budgets, headquarters, a board of directors, a chief executive officer. Schools within a district are much like retail outlets (where direct customer contact occurs). Organizationally, school districts are much the same as a large corporate enterprise. When quality experts talk about things like "working with your suppliers," it is not too difficult to see that, in education, this would mean working with the educational institutions from which students come (their previous schools) or with the parents who send their children. By improving these relationships, we help to improve the quality of students coming into our system.

I believe that there are mostly positive aspects to adapting concepts and words such as 'worker,' 'supplier,' and 'customer' to education. They are valuable terms because they cause us to think about the processes of education in new ways. They help us focus on the performance of our system and to better evaluate its benefit to students, parents, and the community.

Some see potential harm in adopting a model from the competitive, performance oriented world of business where the 'bottom line' is profitability. But we shouldn't forget that businesses are profitable precisely because they offer services or products that society and their customers value. To be profitable in business, one has to pay close attention to what others value. Service organizations, like education, would do well to take much more careful notice of their customers' needs and wants. Our customers include students, parents, employers, institutions of higher education, and others.

Some educators don't like an approach that measures performance in order to improve it. They fear that a method that focuses on performance, both for the students and staff, may increase fear of failure and thereby stifle performance in the long run. But human experience tells us that frequent, precise feedback and assessment of meaningful activity are integral to improvement as well as enjoyment. Any artist, dancer, writer, or athlete who has attained a high level of skill has done so by paying close attention to feedback and frequent evaluations of performance. Although some might disparage the use of data and the emphasis on performance that is present in the quality approach, it is precisely the lack of data and feedback that accounts for the failure of schools to deliberately and systematically learn from their mistakes, build on their successes, and thus continually improve.

Some who are critical of the quality sciences have taken the view that quality is just another new idea, a temporary fad that has captured the imagination of journalists, business gurus, and managers with its high-sounding words and emphasis on data and performance. However, the degree to which quality science has succeeded in transforming American business and industry into internationally competitive organizations speaks loudly for quality science as being much more than just a fad.

QUALITY SAVINGS: REDUCING THE COST OF POOR QUALITY

Poor quality costs money. This is obvious when we think about the cost to consumers or to the producers of poorly manufactured products like clothing, electronics, or automobiles. Producing inferior products or services costs millions in terms of scrap and rework and doing over again things that should have been done right the first time. We also see the cost of poor quality if we are ever on the receiving end of inferior products or service.

Poor quality in education also costs money. Having students repeat courses, providing remedial instruction in schools, teaching high school courses at the college level all cost money, not to mention the social costs to society of producing students who lack necessary job skills. Much of this cost is caused by poor quality in some aspect of the education enterprise.

SCRAP AND RE-WORK

An 8th grade teacher was talking one day to a 7th grade teacher.

"What do you teach?" he asked her.

She named a number of skills and concepts that she covered in her 7th grade classes.

"Well, that's what I teach in my classes, too," he said.

The next day, this same 8th grade teacher was talking to a 9th grade teacher, and he asked, "What do you cover in your classes?"

The 9th grade teacher named what he taught in the 9th grade, and it was the same as what the 8th grade teacher and the 7th grade teacher taught.

"You know," he said, "I suddenly realize that our whole lives are scrap and re-work."

A recent experiment conducted in seven Florida school districts shows the extent to which costs can be saved by applying quality processes to education. These seven school districts initiated some projects to explore the savings that could be accrued through applying quality processes to some aspects of their operations. The seven districts selected their own projects based on the following criteria:

- *Build on success.* The districts were asked to select a process that was clearly in need of improvement. That is, they were asked to select the most obvious targets for improvement.
- *Target 20% savings.* The goal was to get a 20% reduction either in time, money, or number of steps in a process.
- *Begin with a flowchart.* Districts started by developing a flowchart of the process to be improved. After that, each district was free to develop its improvement process independently. Many of them utilized other quality tools, explained in the next chapter.
- *Write procedures for the improved process.* The districts then wrote new procedures for the process and followed the new procedures for a period of six months.

The insert "Projects for Quality Improvement" shows the projects selected by each of the seven districts. Below this are "Results."

The results of the experiments were gratifying. Not every district achieved the same level of process improvement, but each district did achieve a quantifiable success.

PROJECTS FOR QUALITY IMPROVEMENT

District	Process to Be Improved
Dixie	Process for referring a student for a psycho-educational evaluation.
Levy	Process for monitoring absentee rate within the district.
Madison	Process for referring a student for the gifted program or for special education.
Orange	Process for reporting grades in middle and high schools.
Pasco	Process for selecting administrators.
Santa Rosa	Process for complying with new Federal procedures for students with disabilities.
Volusia	Process for handling a parent's request for child to attend a school in a different school zone.

RESULTS

District	Results
Dixie	Cost-savings of $20,000 expected (annually).
Levy	10-step process reduced to 5 steps. Potential cost savings is expected in terms of personnel time.
Madison	20 minutes less time for each referral @ 300 referrals per year.
Orange	Personnel hours savings of $2,600. Savings in supplies & computer time of $532 annually.
Pasco	Personnel hours savings of $7,000 (annually).
Santa Rosa	New process added with no new personnel hours.
Volusia	Personnel hours savings of $23,000 (annually).

—The Center for Educational Technology, "Quality Savings: Reducing the Costs of Poor Quality," **Florida SchoolYear 2000 Technical Report SY-SC11A** (Florida State University, Tallahassee, FL: Spring, 1993).

BENCHMARKING

In the business world, benchmarking is an effective and well used strategy for quality improvement. Businesses are constantly and systematically measuring their performance against that of other similar organizations. They study other high performing organizations closely to determine how they achieve those performance levels. Then they use the information as the basis for their own targets, strategies, and internal programs. Benchmarking is a criterion of the Malcolm Baldrige National Quality Awards application and is an essential tool for quality improvement in the business world.

Schools could well adopt the benchmarking technique as a key element of their quality improvement efforts. "Going to school" on how the best private and public schools achieve successes in certain areas would provide substantial information about what level of performance is really possible and understanding about why a gap may exist between one's current performance and optimum performance.

There's more to benchmarking, however, that just making a few phone calls, attending a conference together where you share information, or even just visiting another school. To do benchmarking right, you have to do a lot of homework first. You have to carefully select a benchmarking partner and determine what you want to find out as a result of the benchmarking process. Typically, benchmarking is done for a very specific reason (i.e., you want to find out how the benchmarking partner does enrollment, or how it uses a specific instructional program). Also, benchmarking can be done between two organizations that may be in a completely different business. Some examples:

- A school might benchmark its inservice training against that done by a large travel agency.
- A public school might benchmark its procedures for teaching science against those of a high performing private school.
- A school might study an organization like the Sylvan Learning Centers or Woodland Hall Academy to compare how they deal with students with specific learning disabilities.

STEPS TO BENCHMARKING

Benchmarking is more than just comparing one organization with another. Rather, benchmarking focuses on a specific area that we wish to study to determine how well we are doing and what level of performance is really possible. Therefore, the first step in benchmarking is to identify an area that is in need of improvement. Again, this area may be quite specific (communication with parents, registration, in-service training, or a specific instructional program).

The next step, after identifying a need for improvement, is to select a benchmarking partner. Selecting the right benchmarking partner is a key step in the process. The idea is to select a partner who is truly exceptional in the area you want to benchmark. The goal of benchmarking is not to show that you are already as good as another organization in an area. The goal is improvement. So you want to observe an organization that is doing a *much* better job in that particular area. You want to pick a partner who will inspire a great deal of improvement and new ideas. One useful question to ask when seeking a benchmarking partner is: "Which organization in the world does the best job of _____?" Remember, a benchmarking partner can be a public school, a private school, a bank, a hospital, a private career college, or some other kind of organization. The goal is to study the best example you possibly can in a very specific area of interest.

The third step is to select a benchmarking team to work on the benchmarking project. Because the team members will likely be unfamiliar with the benchmarking process, they will need to read a good amount of literature about benchmarking techniques. Some of the additional readings cited at the end of this chapter would be appropriate. It is generally recommended that each team member be assigned a specific area of responsibility within the process. For example, if you are going to observe how a benchmarking partner provides a safe school environment, different team members may be assigned areas such as teacher's role, management's role, student's role, communications, budget, and policy. Each team member can then formulate a set of questions about the specific area of interest. These can be compiled and sent ahead of time to the benchmarking partner.

The next step in the benchmarking process is to make a site visit to the benchmarking partner. During the site visit, the team members observe and

---

STEPS TO BENCHMARKING

STEP 1. Identify a need for improvement
STEP 2. Select a benchmarking partner.
STEP 3. Select a benchmarking team.
STEP 4. Read literature on benchmarking techniques.
STEP 5. Assign each member of the team a specific area (enrollment, transportation, safety, discipline).
STEP 6. Have team members formulate questions about their specific areas.
STEP 7. Send compiled questions to benchmarking partner prior to the benchmarking site visit.
STEP 8. Conduct a site visit to the benchmarking partner.
STEP 9. Debrief and plan for action.

ask questions about their specific areas of interest or general questions about the process being benchmarked. Usually, the benchmarking partner will be prepared to respond to the set of questions that were sent in advance as well as to questions asked during the site visit.

The last step is for the benchmarking team to meet together back at its home institution to debrief and share what was learned at the site visit. At this debriefing, each team member can share the particular insights or lessons learned that pertain to that individual's specialty area. Also, plans can now be formulated for making improvements based on what was learned.

THE AT&T UCS BENCHMARKING EXAMPLE

Several public school districts in Florida have been experimenting with the benchmarking process. These districts are part of the Florida Schoolyear 2000 Initiative. The districts chose to benchmark themselves against AT&T Universal Card Services (AT&T UCS) in order to learn how to improve processes which the districts share in common with that company (answering questions on the telephone, training the work force, measuring processes, etc.) The districts were interested in how AT&T UCS provided customer service and how they corrected problems when these occurred.

AT&T UCS was selected as a benchmarking partner for a number of reasons: (1) it operates the Universal Card University, which is their integrated centralized delivery system for education and training; (2) as a service organization, it has many functions that parallel numerous school efforts; (3) it is a past winner of a Malcolm Baldrige National Quality Award, and (4) the company is located in Jacksonville, Florida, so it was easily accessible for the benchmarking visit.

A benchmarking team was selected to work on the benchmarking project. The team members read literature and received some initial training in the benchmarking process. Each member was assigned a specific area to think about in relation to customer service, such as enrollment, transportation, food/meals, school activities, and communication with parents and others. During the site visit, each member was to think about how AT&T UCS customer service might apply to these areas of school operations. Seventy-eight questions were formulated and sent to AT&T UCS in advance of the benchmarking visit. Then, the benchmarking team visited AT&T UCS in its Jacksonville offices.

Results

Most of the questions submitted to AT&T UCS prior to the site visit were answered during the visit. AT&T Universal Card Services operates on the concept of "customer delight." That is, their goal is to do what the customer wants. Its business is 100% service oriented, therefore, customer delight is paramount to gaining and retaining customers. Its theme is "customers are the center of our universe," and they believe they must take the extra step to serve their customers, even though a customer's problem may fall outside their realm of control. This practice builds trust, loyalty, and customer delight. The following section describes processes that AT&T UCS uses in its operations that could be applied in schools.

AT&T UCS Processes that Could and Should Be Used in Education

Screening. At AT&T UCS, front line workers (those who take calls from customers) are called Associates. All Associates are trained and empowered to deal with the majority of incoming calls and the associated questions, problems, or concerns. (This would be like school office staff and district office staff who are authorized to deal with the majority of incoming calls, questions, and concerns or problems from parents or from the community). One of the goals of AT&T UCS is to satisfy customers not only with a single call, but with a single contact during the call. If the Associate cannot answer the question, a memo is sent via computer with all pertinent information to assist the individual receiving the transferred call. If the number of calls not answered on the first contact increases to the point where a trend is emerging, data is gathered on the incoming calls. Calls are sampled and coded in order to determine the training necessary for Associates to complete the call. Often the data collection will include not only a code but also a short narrative to describe more completely the type of calls being transferred.

Such an efficient and effective method of handling incoming calls, problems, questions, and concerns has not been implemented in most schools. A recent survey conducted by TARP of seven Florida school districts revealed that parents who contacted the school or district about a problem or question often had to speak with five or six individuals in order to answer the question or resolve the problem.[2] Even then, most were unsatisfied with the way in which the problem was resolved. To say the least, this type of difficulty in contacting the school or district about a question or resolving problems does not breed high levels of customer satisfaction.

Resolution Not Escalation. When a customer insists on speaking with the Chief Executive Officer, front line Associates offer to transfer the call to a specialist or manager. If the customer is insistent, the call goes to the Executive Response Center, where customers can speak to a representative of the president's office. These representatives can usually resolve the problem. The goal is to do what is appropriate to get the problem solved to the customer's satisfaction rather than needing to escalate the call to the highest level.

Understanding and Using Quality Measures. AT&T UCS collects data on over 100 quality measures every day. Measurement and management are synonymous at AT&T UCS. They believe that if a process cannot be measured, it cannot be improved. Measurement is dynamic and changes frequently. They find that financial measures combined with quality measures yield successful results. The measures are gathered daily and discussed daily. Workers do not just compile data to sit on a shelf unused. The discussions of the data are as important as the measurements themselves, as this step initiates the action necessary for improvement. And the company acts on the information gathered. Any area identified as needing improvement is acted on within 24 hours.

The AT&T UCS operations management meets at 9 o'clock every morning to review the data from the previous day. It reviews 100 measures on 104 processes. The results from all measures on all processes are displayed each day in the entrance to the corporate headquarters.

Measurement in Simple Terms. The measures themselves are easy for all to understand. All 100 measures are grouped under four easy to understand names. Each unit uses the appropriate measures and reports each day on its time, accuracy, availability, and courtesy. In education, the set of measures might be different, but the idea of a small set of easy to understand indicators of quality would be helpful to schools.

Names of Measures	Examples
Time	Time taken to prepare statements
Accuracy	Accuracy of receipts
Availability	Availability of data
Courtesy	Courtesy with customers

A Variety of Measure to Indicate Quality. A variety of data collection methods are used for the measures, which are reviewed daily. Measurement methods include: call monitoring, contracted surveys, suggestions and complaints, focus groups, customer advocates, customer panels, attrition reports, and reasons for contacting the company. A minimum of 50 calls per day are monitored, and the content is analyzed to measure performance and to identify any developing trends. Performance measures are changed constantly to reflect emerging trends identified in the daily measures.

Understanding Customer Satisfaction Data. At AT&T UCS, data are collected, analyzed, and discussed on a daily basis. Cross functional teams are employed to identify key attributes for customer satisfaction. Cross functional teams are made up of individuals from various department and different levels within the organization which help with communication between departments. Finally, all measures are "wrapped" into a single company reliability score which is reported daily and displayed throughout the building. Every day, the entire organization has the opportunity to earn additional compensation by meeting or exceeding the standard for 96% or more of the daily indicators.

Recognition. Over 40 recognition awards at AT&T UCS are given, mainly through peer recognition. They believe it is better to give too much recognition than too little.

Failures. Failures are celebrated at AT&T UCS. They are celebrated because each failure is viewed as an opportunity to improve. And *workers take specific action on each failure* to help insure that it won't happen again. Historically, this has not been done in schools or government because admitting failure has been unacceptable—thus the lack of willingness to take action. Celebrating failures focuses on the opportunity to learn and to improve processes.

Assuring Corrective Action. Assuring effective corrective action requires measures and the review of the measures to confirm lasting improvement. Corrective action should be proactive rather than reactive.

Special Education. AT&T UCS has a special way of dealing with employees or customers with special disabilities or needs. The company relies on input from the individuals themselves rather than deciding what is best for them. Furniture and physical work environment are designed to meet the needs of the individuals involved.

Satisfied Employees Satisfy Customers. Employee satisfaction is a priority at AT&T UCS and, as such, is also measured. Responses by employees to a survey are compared with employee responses from 600 other companies to the same survey. AT&T UCS has performed better than both the national norms and high performing norms. Still, the company sees this as an opportunity to improve.

Suggestions for Schools

At the conclusion of the benchmarking site visit to AT&T UCS, the people there offered suggestions for customer service and corrective action in schools.

1. Find Out Why Callers Want to Speak with the Principal. The first suggestion was to identify why callers insist on speaking with the principal. In the opinion of AT&T UCS, callers may feel that the principal is the only individual capable of answering their question or resolving their problem.

Quality Science 109

> **QUALITY IN A CRISIS**
>
> Four years ago, things were not working out well at AT&T Universal Card Services. Applications were not being processed fast enough, they ran out of plastic for new cards, and monthly billings were not getting to customers on time. Family members were working night and day to help meet the overload of work. The CEO and his son pitched in to answer customers' calls.
>
> One day, during this critical period, the CEO asked where the quality staff was. When he was told they too were pitching in during the crises, he said, "If we don't pull them off the line operations and put them to work on our quality system, we will never get out of this crisis." (This is in contrast to some schools where they may not have a quality work group or where their quality work group members have been given additional line responsibilities to meet a current school crisis.)
>
> —*Technical Assistance Research Programs,* **Customer Dissatisfaction Surveys for School Districts: Overview of Parent/Guardian Dissatisfaction with Feeder Pattern Schools in Florida Schoolyear 2000 Sites** *(Arlington, VA: December 7, 1994).*

A related suggestion was to gather data on incoming calls and do a Pareto analysis (see Chapter 7). Following the analysis, train the office staff to handle the most frequently occurring types of calls.

2. Find Out If Parents Know Who to Contact. A related suggestion was to survey the parents of students to identify if they know who they should contact at the school or district for various kinds of questions or problems—vice principals, counselors, teachers, office staff, etc. If the survey reveals little understanding or much misunderstanding, provide information to the parents about the roles of various school and district personnel.

3. Use a Continuous Quality Improvement Model. The fourth suggestion was that schools should operate under a customer-supplier model. View the students and parents as customers and the school as the supplier. Make efforts to insure that the needs and wants of each customer is well understood. Use daily measurements to show if corrective action is effective both in the short and in the long term. Talking to previously upset or unsatisfied customers identifies way to prevent future failure.

4. Establish a Customer Service Center. The fourth suggestion was to establish a customer service center in each school. According to AT&T UCS, a portion of customer service means making efforts to correct the process that required a customer service function in the first place. The goal of a special customer service center is, ultimately, to eliminate the need for such a center. While realistically, a customer service center may be required, efforts should be concentrated on improving the root causes of the problems that cause people to need to contact the customer service center in the first place.

5. Resolve Problems in the Customer's Favor. The fifth suggestion is to encourage exceptions to policy or practice which resolve problems in the customer's favor. Making exceptions to policy should not be considered failure.

NEW ROLE OF THE CUSTOMER

Reformers are looking seriously at the importance of the role of the customer. Choice and voucher programs, which allow students to take their education dollars anywhere they choose, continue to surface in the public debate. The potential positive effect of free enterprise and competition on a bureaucratized and often beleaguered public school system may be very powerful. Also, the creation of new kinds of private schools that will deliver a new form of education will likely put pressure on public schools to improve their quality. Even beyond this is the ever increasing potential of delivering instruction to learners at home or other locations. As global computer networks and telecommunication channels thrive, learners will have increasingly more access to learning resources targeted at their particular interests, pace, and learning styles. Computers that offer information and help at the time of need are already providing "just-in-time" learning to a growing number of auto mechanics, power plant operators, managers, and other workers in banks, businesses, law offices, and other settings. "Just-in-time" learning means that the person can access exactly the information, help, or resources needed, quickly and easily (usually over a computer terminal or portable device). As these automated learning resources continue to increase, learning will no longer be confined to the box of a classroom. Instead, learning and learning resources will reach students outside of school. The most important implication of these changes is that they represent a fundamental shift in power from education institutions to the consumers—students and parents. As consumers, students and their parents will have more opportunity to select and choose educational products and services that respond directly to their personal demands. It is the changing role of the customer that has made quality an issue of survival for American business. It will become just as critical for education.

VIRTUAL SCHOOLS

The corporate world is talking a great deal now about "virtual corporations." Virtual corporations are formed when marketing, manufacturing, and servicing companies come together as needs arise and dissolve when needs are satisfied. Virtual corporations foster the most fitting use of resources, especially human resources. High-speed electronic networks that didn't exist a few years ago make these temporary alliances possible. These electronic webs of coordination and cooperation spread across the planet, from one company's computer network to another's, creating virtual corporations that meet a particular need at a particular time.

To meet increased and varied customer needs, the school may indeed have to become more of a *virtual school*—functional and purposeful but with less emphasis on buildings or a specific location. The phrase *virtual school* is used because this entity possesses the powers and capabilities of a school but is not in the form of a school as we know it today. It resembles a school in its functions and purpose, but it is much more versatile in that it can exist anytime or anywhere. The virtual school would be a dispenser or source of information, knowledge, learning resources, and methodologies, but it would deliver instruction to learners in various locations at various times throughout the day and year—home, learning centers, workplace, care centers, etc. Rather than a school in the traditional sense, a virtual school is more like a spider web or a network with a small central core that coordinates the various activities and adds value.

A virtual school might be funded by public or private monies or by a combination of sources. But to the outside observer it will appear almost edgeless, with flexible and continuously changing relationships between the school, public and private learning resources,[3] cable TV, satellite, and telephone companies, and students. Its function is to give the students exactly what they need in the most effective form for them to learn and use it. Schools with buildings will continue to exist and to serve useful child care and social functions, but educational services may be delivered through a combination of other means.

One of the keys to designing virtual schools will be the development of enabling technologies to support learning anywhere, anytime. These new technologies will need to maintain integrated and ever-changing data files showing individual student progress on various lessons, and they will need to provide students with an array of well developed learning resources and instructional methodologies. That means that someone must develop new and more sophisticated learning management systems that enlist and empower the student in the design of the learning process itself. Through the use of sophisticated learning and advisement software, the student can take on responsibilities for himself or herself that were heretofore reserved for the teacher. Students could also have access to sophisticated information networks that gather data on the requirements of employers and colleges and then analyze what the student's deficiencies are in terms of those requirements. Then the systems could allow the student to correct those deficiencies by using the instructional methods that are most appropriate for that student.

VIRTUAL PRODUCTS

Hand in hand with the idea of a virtual school is the idea of a *virtual product*. A virtual product or service is one that is produced instantaneously and is customized in response to customer demand—products or services that are created at the time of need and uniquely fashioned for a specific customer.

Examples of virtual products are springing up all around us as businesses that serve people strive to customize their services to specific, individual customer needs:

- Prescription eyeglass lenses are ground and placed in custom frames within 60 minutes by companies like Lenscrafters and Pearle Vision Express.
- Levi's jeans now provides made to order Levi's for women. The store clerk takes the woman's measurements and zaps a digital blue-jeans blueprint to a Levi's factory in Tennessee, where a robotic tailor produces the custom pair of jeans.
- Taco Bell instituted less-than-20-seconds service at indoor counters and drive-through windows.
- Polaroid gave us 60 second photography years ago, but even that has been largely surpassed by 1-hour developing and printing of high quality conventional photographs.
- Electronic cameras play pictures on a TV set a moment after they have been taken.
- Personal computers and laser printers have made instant desk-top publishing a reality in millions of offices and homes around the world.
- Oil is changed in cars in ten minutes.
- Travel reservations are made with electronic speed.
- We can obtain cash instantly at ATMs.[4]

Should education follow the lead of these customer service organizations by creating virtual products and services as a way to better meet the demands and preferences of their customers? That is a question education innovators will need to answer. But it seems highly likely that unless we do so, schools will be viewed as antiquated oddities in a world of cost-effective, instantaneous production of customized services.

In a customer oriented environment, good decisions can be made on the spot. Because of the availability of computers that can gather and analyze information and present it at time of need, customers (students) and a teacher dealing directly with a student can make a decision and implement that decision immediately. Even less motivated students can be turned-on by higher quality instruction that is designed for their particular learning needs, providing opportunities to succeed in areas where they failed in the past. In other words, the system can be responsive to the individual needs of the customer. Any conceptualization of restructuring education that does not take this new customer empowerment into account is merely attempting to streamline the past.

The advent of computers, new forms of organization, electronic learning management systems, electronic performance support systems, and numerous other technological advances have made it possible for companies to be market responsive and cost-effective at the same time. Flexibility and responsiveness are now keys to competitiveness. The same will become true for the education enterprise. The emphasis on quality and customer satisfaction will become less and less discretionary. Those schools that delay or refuse the process of the third technology—quality science—will be swept away, their remnants seized, reorganized properly, and absorbed by fast-moving modern competitors.

MEASURING CUSTOMER SATISFACTION

One of the principal challenges in applying quality science to education is that of measuring customer satisfaction. Obtaining the customers' point of view about the effectiveness of the organization's products and services is basic to any application of quality science. The purpose of customer satisfaction measures is to provide managers with the customers' point of view about how well it services its customers. Unfortunately, in education, we have yet to apply known methods for measuring the level of satisfaction of our customers.

Measuring customer satisfaction involves a process that is much more scientific, rigorous, and extensive than the typical parent surveys sent home today by most schools and districts. Most of these simply ask people to rate different aspects of the school on a scale ranging from "very satisfied" to "very dissatisfied." Such surveys don't even come close to fulfilling the critical role of customer satisfaction measures in quality science. That's because they don't identify where processes fail or why. They measure people's general *attitudes* about the organization rather than obtaining information about specific problem experiences or transactions, and they do

MOST PARENTS DON'T COMPLAIN

Most parents don't complain. As shown in the diagram below, only 5% of parents ever complain to district headquarters. Only 39% complain to a school administrator. Fifty-six percent encounter problems but never complain to anyone in the school system. If your school or school district relies on the quantity of comments or complaints that you receive at the school or district office as an indicator of the level of quality in your organization, you need to realize that you are seeing just the tip of the iceberg. Most school districts highly underestimate how dissatisfied their customers are.

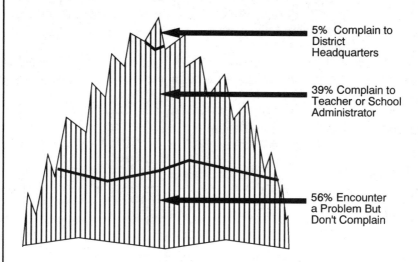

5% Complain to District Headquarters

39% Complain to Teacher or School Administrator

56% Encounter a Problem But Don't Complain

Why don't more parents complain? When parents who had a problem but did not complain to anyone are asked why, these are the typical answers:
- Wouldn't resolve question/concern
- Not worth time or trouble
- Don't know how or whom to contact
- Didn't want to upset anyone or get someone in trouble
- Tried but couldn't get through
- Fear of retribution from the criticized administrators or teachers

All of these are commonly cited reasons for parents who do not complain. Fear of retribution is fear of face to face encounters, or more specifically, fear that a teacher or principal might penalize the student in some way if the parent were to complain.

—Adapted from **Technical Assistance Research Programs, Customer-Driven Quality Workshop,** *Florida State University, Center for Educational Technology, Florida Schoolyear 2000, February, 1994,* and **Technical Assistance Research Programs, Customer Dissatisfaction Surveys for School Districts: Overview of Parent/Guardian Dissatisfaction with Feeder Pattern Schools in Florida Schoolyear 2000 Sites** *(Arlington, VA: December 7, 1994).*

Quality Science

not measure the frequency and type of customer complaints or inquiries or the adequacy of the responses made by the school or district to those complaints or questions. Most of the surveys used in schools today make it look like the customers don't see any problems with the school or district—no complaints. But we found that if you go out and ask the customers directly, "Did you have a problem, question, or complaint in the last six months?" most of them did. And they will be happy to tell you about it on a survey once you've asked the question very specifically.

Schools reaping the least benefit from their satisfaction measurement systems are those that conduct satisfaction surveys to convince the school board that most parents are completely satisfied. Surveys designed to validate happiness have little benefit. Either the survey will show that everyone is happy, in which case, no action will be taken, or the survey will reveal that no one is happy; and because there is no information about what to do to correct the problem, no action will be taken either. There is really no way to improve the quality of service other than to identify those specific areas where customers' expectations are not being met.

A quality science approach uses surveys and questionnaires sent home to parents and other customers, but in this case, the questionnaires are designed to measure specific instances of *dissatisfaction* among a specific customer group. For example, one questionnaire may be designed for parents, another for students, and a third for the business community. These questionnaires request very specific information about the number and types of concerns or questions that customers have experienced with the school or school district. In addition:

- The questionnaires examine customer contact behaviors—Did they contact anyone? If so, who? Were they satisfied with the action taken? Were they satisfied with the response? With the personnel?
- A true customer satisfaction process identifies the "market impact" of these problems. That is, responses are used to estimate, to the degree possible, the number of parents who would switch to another school or district if they had a choice or who will pass along negative word of mouth comments to their neighbors and friends as a result of any particular problem or set of problems.
- The process identifies the key drivers of satisfaction—those areas that have the most significant positive or negative impact on customer loyalty.
- The process involves "mystery shopper" telephone calls to various schools across the district to determine the degree of consistency and effectiveness in responses to routine community questions or concerns.
- The customer satisfaction process compares the school or school district against other service industries within the public or private

sector. These might be private schools, government agencies, employment services, etc. This process, described earlier, is called *benchmarking*. It compares the school services with "best in class" businesses, agencies, and organizations that are similar to schools (child care centers, private schools, hospitals, human service agencies). This comparison against other, similar organizations provides a point of reference—a standard against which the school can compare itself. Benchmarking is important because it tells us what level of performance is really possible and helps us understand why our current performance may not be at that optimum level.

The effort to measure customer satisfaction must be part of a total quality system that sets customer-driven priorities for improvement and positions the school or district to move from a *fire fighting* into a preventive mode of service.

Information from the customer satisfaction process is essential to improvement. In a quality science approach, the school district must develop a district-wide communication plan that ensures consistent, effective, and timely responses to customer questions and concerns by providing district and school personnel with the necessary knowledge and expertise for responding to customer questions and concerns. The system must aggressively *solicit* customer questions and concerns on an ongoing basis *and* ensure responsiveness to those concerns. In most businesses, this requires setting up a customer assistance process at each location to ensure responsiveness to local customers and putting in a telephone *hotline* to provide outreach to customers seeking additional assistance. This would be a basic step in the quality science approach for any school or district. The hotline does not replace personalized service at the district or school but serves as a *one-stop shopping* channel for routine questions, confused customers, or customers seeking assistance after they have gone through routine channels.

We recently piloted such a customer satisfaction process in one district in Florida and are now in the process of expanding it into seven additional Florida districts. In the pilot district, specially designed surveys were sent to parents and representative members of the business community. The questionnaires requested specific information about problems or concerns that the person may have had in relation to the school or district in the last six months. They asked about the person's "contact behaviors"—did he or she contact anyone about the problem? If so, who? What was the result? Were they satisfied with the action taken? With the personnel? Why or why not? The survey was also used to estimate the percent of customers potentially lost due to specific types of problems or complaints. This was done by asking the people filling out the survey how likely they were to not recommend the school or district because of the problem or complaint and if they would definitely/probably withdraw from the school or district if given a

> **SAMPLE QUESTIONS FROM A CUSTOMER SATISFACTION SURVEY**
>
> The survey first asks the person to check areas of his or her most serious concerns from a list of choices that the school or district has determined are frequent areas of questions or concerns. Then it asks:
>
> Did you contact anyone about your MOST SERIOUS question or concern?
>   1. Yes   2. No
>
> Why didn't you contact anyone? (CIRCLE ALL THAT APPLY)
> 1. The question/concern wasn't important enough to make contact
> 2. I didn't believe contacting would resolve my question/concern
> 3. I didn't know how or whom to contact.
> 4. The teacher was not available
> 5. The principal was not available
> 6. I didn't want to upset anyone/get anyone in trouble
> 7. I tried, but couldn't get through to anyone
> 8. Other (PLEASE SPECIFY): _____
>
> When you made a contact, how many different people did you have to speak with or how many times were you transferred before reaching someone who could assist you?
>
> How satisfied were you with the person who handled your contact in EACH of the following areas? [these were listed]
>
> How long did it take from the time you FIRST contacted someone until FINAL action was taken to answer your question or resolve your concern?
>   —*Customer Satisfaction Survey*, Technical Assistance Research Programs and Center for Educational Technology, Tallahassee, FL: Florida State University (July, 1993).

choice. "Mystery shopper" telephone calls were placed to various schools in the district to determine if there was consistency in responses to routine questions. Lastly, the level of customer satisfaction with the schools was compared with the level of customer satisfaction of similar types of organizations. In this case, the school district came out among the lowest in terms of customer satisfaction and loyalty in comparison to people's satisfaction with private schools or other government, health, or human service agencies.

As a result of this pilot, we found that existing communication policies and practices in the district and schools were the major cause of customer dissatisfaction with that district. Also, almost all of the parents and business respondents that indicated that they did not contact the district or the schools about their most serious concerns said that they *would have* contacted someone if a special hotline were available. After seeing the results of the survey, the district did implement a customer assistance hotline. The hotline was prominently advertised in school publications and in local newspapers and on television. Trained school personnel, volunteers, and school district retirees were used to handle the incoming calls. Due to budget constraints, the hotline was operated only during "crises" times during the school year (two weeks before and after registration, two weeks before and after mid-year, and two weeks before and after the end of school.) The district has also established a centralized customer communications assistance protocol (rules, manners, etiquette, decorum, and procedures) for handling and resolving customer problems and complaints.

WHY IT PAYS TO PROVIDE CUSTOMER SERVICE

Coca Cola, Chevrolet, and American Express were among the early companies that decided to invest millions of dollars in the 800 customer service hotlines that we take for granted today. Why were they willing to invest this amount of money to provide customer service? The reason is very simple: Customer satisfaction produces customer loyalty. And customer loyalty means increased profits for these companies.

Is customer loyalty important for schools? It should be! Schools (it doesn't matter whether they are public or private) depend on their customers for their ultimate existence. It is their customers that provide tuition in the case of private schools and the tax base for public schools. Customer support, customer satisfaction, word of mouth comments, newspaper articles, letters to the editor, and parent support all have an impact on the school system. Of course, these will have a much greater impact in districts that are moving toward a free market type environment where students and parents choose which schools to attend. But even in districts with attendance zones, the impact of high or low quality customer service is felt—superintendents lose their jobs, school board members are voted out of office, and home buyers shop for homes in the best school zones. The bottom line

---

CUSTOMER SATISFACTION: A DEFINITION

The state in which customer needs, wants, and expectations throughout the service's life are met or exceeded, resulting in repeat services and loyalty, and positive word-of-mouth behavior.
—*TARP, Customer-Driven Quality Workshop,* Florida State University, Center for Educational Technology, Florida Schoolyear 2000 Initiative (February, 1994).

in customer satisfaction is the customer's willingness to continue to support the relationship.

Research on customer satisfaction has shown that bad word of mouth comments typically reach twice as many people as positive word of mouth comments. That is, when consumers have a bad experience with an organization, they will tell 10 people about the bad experience. They will only tell only five people about a good experience.

CUSTOMER DRIVEN QUALITY

Customer driven quality means that we take extreme care to understand the needs, wants, and expectations of our customers. It also means that we take extreme care to find out which specific expectations we failed to meet and on which we will experience the most disloyalty or negative word-of-mouth comments. The goal is to identify those areas where we failed to meet expectations. To achieve high customer satisfaction, it is not necessary to delight our customers on every single item or service, because we can't possibly do that. What *is* necessary is to meet or exceed our customers' expectations in those areas that are key in the customer's mind. If we try to make *everything* a high priority for quality, nothing is a high priority. We want to designate as high priority those services, items, or interactions with customers that are "points of pain" to the customers. "Points of pain" are those interactions that cause the highest level of concern to the customers. These items represent "moments of truth" in terms of customer satisfaction because customers will judge the quality of the entire system by the quality of these critical interactions. It is much more important to be consistently excellent in these "moments of truth" situations than to be occasionally excellent in many other areas.

CUSTOMER DRIVEN QUALITY

1. Which products or services are most important to the customers?
2. What are the processes that produce these products and services?
3. Which processes have the highest visibility with customers?
4. Which processes have the greatest impact on customer-driven performance standards?
5. Which processes do performance data suggest have the greatest potential for improvement?

—Adapted from Arther R. Tenner and Irving J. DeToro, **Total Quality Management: Three Steps to Continuous Improvement** *(Reading, MA: Addison-Wesley, 1992).*

PLANNING FOR CORRECTIVE ACTION

In our pilot test of a customer satisfaction process in Florida, we found that there is little action taken on the information that schools and districts get from their customers. In most of the school districts we examined, 95% of the resources that go to customer service are used for responding to the same problems, complaints, or questions over and over again. Only five percent of the resources were devoted to the activity of taking the information that schools or districts have about customer complaints or questions, analyzing it, feeding it back to the rest of the system, identifying the root causes of the problems, and implementing meaningful and lasting changes to prevent the same problems or questions from occurring again and again. Our study in Florida showed that of the customers who did complain, 70% were dissatisfied with action taken by the school or district in response to their complaint.

Some customers (parents, students, taxpayers, employers, etc.) may have unrealistic expectations or misconceptions about the schools or services. When these misconceptions arise, the organization must take action to address them constructively. The customer may not always be right, but the organization will still suffer the consequences of dissatisfaction, discontent, and negative word-of-mouth in the marketplace unless positive steps are taken to educate the customers about the service or how to use the service.

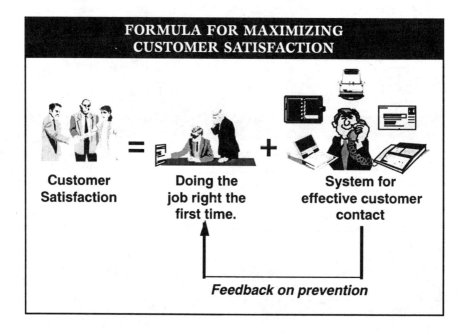

This problem (customer's unrealistic expectations or misconceptions) occurs regularly in business. People mistreat or misuse a product, patients don't properly take the medicine that the doctor prescribed, clients misunderstand or fail to act on the information or advice they were given. What is called for in these cases is clearer instructions, increased communication, telephone "hotlines," or customer education. Ultimately, whether the customer is right or wrong, the business or service will suffer the consequences.

One positive note: taking positive action on customer complaints or problems has a high effect on producing high customer loyalty. Our customer satisfaction research in Florida confirms this. Customers who complained and who felt that their problems were handled even moderately well showed increased loyalty. That is, they were likely to recommend the school or district to a friend, or pass on positive word-of-mouth comments.

GETTING HELP

Setting up a quality system, measuring customer satisfaction in the way I have described, establishing a corrective action process, and benchmarking quality against other similar organizations are not familiar tasks to most schools and school districts. To do any of these things requires specialized skills and, in most cases, outside expertise. Fortunately, there are numerous sources to which schools or districts can turn for help. Most large businesses (and many small businesses) have well established quality programs that include proven procedures for measuring customer satisfaction. The larger companies may have a quality department with quality managers, quality specialists, quality consultants, and customer service specialists on staff. Forming a partnership with a local company can be the most efficient means to tap this expertise. Participating business partners may allow their employees to contribute their time to help at a technical level. In addition, several new national consulting groups have emerged to assist school districts to establish customer satisfaction processes and quality systems. Technical Assistance Research Programs (TARP) of Arlington, Virginia developed the excellent customer-driven approach to school improvement that we use in the Florida schools I mentioned earlier in the chapter. Westat, Inc. of Rockville, Maryland also works with schools and school districts to help them set up quality processes. In addition, The American Society for Quality Control (ASQC), the largest organization in the United States devoted to quality, recently formed an education division for assisting schools that want to use quality science methods. The ASQC Koalaty Kid Alliance was formed in 1994 and provides a newsletter, conferences, and consulting to schools.[5] The American Association of School Administrators also has a Total Quality Network. The Network offers on-line information services about quality efforts in schools, a bimonthly newsletter, a resource binder, as well as other services.[6]

RULES FOR QUALITY PREVENTION IN EDUCATION

This humorous (only because it is painfully true) observation of efforts to implement quality science in schools appeared in a recent education newsletter:

Recently, citizen groups have attempted to bring quality into education. We professional educators have dealt successfully with these troublemakers before and together we can do it again. The following steps toward Total Quality Prevention (TQP) have served us well in the past and show great promise for the future.

- The first step is to delegate quality at all levels. High level administrators need only to appoint quality assistants to send a clear signal that quality is best done by someone else. Teachers can then delegate quality to their students.
- Next, create a universally accepted vision written by a committee representing all interests in education. A consensus will then form around the status quo, finger-pointing circles will be formed, and accountability will be rendered as harmless as quality itself.
- Finally, form quality committees. It is best to call these committees "teams"—being careful not to give them any specific tasks or problems to solve. They should discuss the many purposes of education, the need for more money, the barriers to quality, and, most critical of all, problems whose solutions lie beyond their control. This activity should continue until everyone is convinced that quality, while important, is really someone else's responsibility.

These three steps have proven effective for years, but consume time. There are some rules for maintaining quality prevention that do not involve everyone. The most effective ones are presented here with the understanding that they are neither exhaustive nor original, but have stood the test of time.

1. When society points its finger at the critical quality education issues of the day, study the finger.
2. Deliver all advice on quality education in pristine condition—never having used it yourself.
3. Point out the differences between education and industry, thus avoiding quality for as long as the American automobile industry did.
4. Point out to people, who may not have noticed, that philosophically there are neither red nor white beads, thus postponing their understanding of variation indefinitely.
5. Redefine Deming's teachings to fit education, being sure to eliminate any traces of statistical reasoning, a systems approach, or measurement of anyone besides the student.
6. Define the teacher as the customer, thus preventing interference by parents, community leaders, and employers.
7. Strengthen the anti-quality vaccine by setting a numerical goal for standardized test scores and holding the students accountable.

> 8. Make changes after studying worthy goals and proven results from other schools—but do not use their processes.
> 9. Avoid confusion by concentrating only on what you do well, thus inoculating teachers against any preoccupation with improvement.
> 10. Commit to solving the needs of the minority, handicapped, and poor students by continuously requesting more money.
>
> A flexible focus on proven quality prevention techniques is vital. Regularly change slogans. Promise quick solutions often. Develop many ways to motivate teachers to work harder. Delegate quality frequently. Maintain schedules and vary performance standards. Save tax dollars regardless of cost. These actions may not prevent quality totally, but it will be on its way out.
>
> WHAT WE REALLY NEED IS A
> NATIONAL EDUCATIONAL QUALITY PREVENTION WEEK!
>
> —F. Craig Johnson., *Instructional Systems News* (Fall 1993).

An ambitious school or district may wish to apply for a state or national quality award. The application preparation process itself is very helpful in that it provides motivation, guidelines, steps, and audits along the way. A Malcolm Baldrige Quality Award is to be given to a school for the first time starting in 1996. Criteria for educational institutions that wish to receive the award can be obtained from the National Institute of Standards and Technology (NIST).[7] Also, most states have quality award programs. A first step might be to check with your local or state Chamber of Commerce.

SUMMARY

Quality science allows us to meet the needs of those we serve—our customers. Education has many customers: students, parents, employers, institutions of higher education, taxpayers, etc. By applying quality science, the third technology, we can identify who those customers are and what they want and expect from what we do day-to-day in education, and then provide that in a quality way—every time. Using quality science in schools is common sense. The quality science approach not only provides the best long-term result, but also allows us to maximize resources and provide the best possible education for children at the lowest possible cost.

Using quality science in education is more than just common sense. It is absolutely essential for effective education reform. But to be effective, it must be combined, harmoniously, with the other four technologies discussed in this book. Efforts to apply quality science can easily clash with traditional methods and old ways of thinking. This is why change management, the fourth technology, is also critical. Change management methods and principles are fundamental keys that must be expertly applied in any effort to introduce quality science methods into an organization, school, or school district.

FOR FURTHER READING

Quality Science

Dennis R. Arter, "Evaluate Standards and Improve Performance with a Quality Audit," *Quality Progress* (September, 1989), pp. 41–43.

Karen Bemowski, "The Benchmarking Bandwagon," *Quality Progress* (January, 1991), pp. 19–21.

Scott M. Broetzmann, John Kemp, and Mathieu Rossano, "Moving from Measuring to Managing Customer Satisfaction," *Customer Service Management* (March, 1995), pp. 16–18.

Frank Caplan, *The Quality System: A Sourcebook for Managers and Engineers*, 2nd ed. (Radnor, PA: Chilton, 1990).

W. Edwards Deming, *Out of the Crisis* (Cambridge, MA: MIT Center for Advanced Engineering Study, 1982).

Christopher J. Fay, "Royalties from Loyalties," *Journal of Business Strategy*, 15, No. 2 (March/April, 1994), pp. 47–51.

Bob Filipczak, "Customer Education (Some Assembly Required)," *Training*, 28, No. 12 (December, 1991), pp. 31–35.

John Goodman, Scott M. Broetzmann, and Colin Adamson, "Ineffective—That's the Problem with Customer Satisfaction Surveys," *Quality Progress*, 25, No. 5 (May, 1992), pp. 35–38.

Robert Janson, "How Reengineering Transforms Organizations to Satisfy Customers," *National Productivity Review*, 12, No. 1 (Winter, 1992/93), pp. 45–53.

Robin L. Lawton, "Creating a Customer-Centered Culture in Service Industries," *Quality Progress*, 24, No. 6 (September, 1991), pp. 69–72.

Henry L. Lefevre, "Variation in the Service Sector," *Quality Progress* (December, 1990), pp. 60–62.

Joseph Oberle, "Quality Gurus: The Men and Their Message," *Training* (January, 1990), pp. 47–52.

Walter A. Shewhart, *Economic Control of Quality of Manufactured Product* (Milwaukee, WI: Quality Press, 1980).

Quality Science and Education

American Association of School Administrators, *Quality Goes to School: Readings on Quality Management in Education* (Arlington, VA: 1994).

American Society for Quality Control, *KoalaTymes* Newsletter. Available from ASQC, 611 East Wisconsin Ave., P.O. Box 3005, Milwaukee, WI 53201–3005.

Susanne Axland, "Looking for a Quality Education?" *Quality Progress*, 24, No. 10 (October, 1991), pp. 61–72.

David L. Bayless, Gabriel A. Massaro, Elizabeth Bailey, David Coley, Royce Holladay and Dana McDonald, "The Quality Improvement Management Approach as

Implemented in a Middle School," *Journal of Personnel Evaluation in Education,* 6, No. 2 (December, 1992), pp. 191–209.

John Jay Bonstingl, *Schools of Quality: An Introduction to Total Quality Management in Education* (Alexandria, VA: Association for Supervision and Curriculum Development, 1992).

Robert K. Branson and Terrelle Buckner, "Quality Applications to the Classroom of Tomorrow," *Educational Technology,* 35, No. 3 (May–June, 1995), pp. 19–22.

Margaret A. Byrnes, Robert A. Cornesky, and Lawrence W. Byrnes, *The Quality Teacher: Implementing Total Quality in the Classroom* (Bunnell, FL: Cornesky and Associates, 1992).

Maury Cotter and Daniel Seymour, *Kidgets* (Milwaukee, WI: American Society for Quality Control, 1993).

Joseph C. Fields, *Total Quality for Schools: A Suggestion for American Education* (Milwaukee, WI: American Society for Quality Control, 1993).

Joseph C. Fields, *Total Quality for Schools: A Guide for Implementation* (Milwaukee, WI: American Society for Quality Control, 1994).

J. Harris, S. Hillenmeyer, and James V. Foran, *Quality Assurance for Private Career Schools* (Washington, DC: The Association of Independent Colleges and Schools, 1989).

Janice L. Herman and Jerry J. Herman, "Total Quality Management (TQM) for Education," *Educational Technology,* 35, No. 3 (May–June, 1995), pp. 14–18.

Dean L. Hubbard, *Continuous Quality Improvement: Making the Transition to Education* (Maryvale, MO: Prescott Publishing, 1993).

Roger Kaufman and William Swart, *Educational Technology, Special Issue on Quality Management in Education,* 35, No. 3 (May–June, 1995).

Roger Kaufman and Doug Zahn, *Quality Management Plus: The Continuous Improvement of Education* (Newbury Park, CA: Corwin, 1993).

David Leigh, *Total Quality Management for Educators: A Ten Part Series* (Temple, TX: Temple Jr. College, TQM/Tech Prep Group, 1993).

Betty L. McCormick (Ed.), *Quality & Education: Critical Linkages* (Princeton, NJ: Eye on Education, 1993).

Lynn Olson, "Schools Getting Swept Up in Current of Business' 'Quality' Movement," *Education Week* (March 11, 1992), pp. 1, 25–26.

Gray Rinehart, *Quality Education* (Milwaukee, WI: American Society for Quality Control, 1993).

Peggy Siegel and Sandra Byrne, *Using Quality to Redesign School Systems* (San Francisco, CA: Jossey-Bass, 1994).

Stanley J. Spanbauer, *A Quality System for Education* (Milwaukee, WI: American Society for Quality Control, 1992).

Jacob O. Stampen, "Improving the Quality of Education: W. Edwards Deming and Effective Schools," *Contemporary Education Review,* 3, No. 3 (Winter, 1987).

Technical Assistance Research Programs, *Customer Dissatisfaction Surveys for School Districts: Overview of Parent/Guardian Dissatisfaction with Feeder Pattern Schools in Florida Schoolyear 2000 Sites* (Arlington, VA: TARP, December 7, 1994). Available from TARP, 1600 Wilson Boulevard, Suite 1400, Arlington, VA 22209; tel. (703) 524-1456.

United States Department of Commerce, *Pilot Education Criteria for the Malcolm Baldrige National Quality Award* (Washington, DC: GPO, December 16, 1994).

NOTES

1. See A. V. Feigenbaum, *Total Quality Control* (New York: McGraw-Hill, 1961) for a discussion of a "system" approach to achieving quality.
2. Technical Assistance Research Programs, *Customer Dissatisfaction Surveys for School Districts: Overview of Parent/Guardian Dissatisfaction with Feeder Pattern Schools in Florida Schoolyear 2000 Sites* (Arlington, VA: TARP, December 7, 1994).
3. These public and private learning resources would include things like on-line electronic libraries, on-line encyclopedias, learning games, on-line lessons, etc.
4. These examples of virtual products are taken from William H. Davidow and M. S. Malone, *The Virtual Corporation: Structuring and Revitalizing the Corporation for the 21st Century* (New York: HarperCollins, 1992), p. 4. The information on Levi's made-to-order jeans is from the *New York Times* and *Glamour* magazines, reported in the *Tallahassee Democrat,* November 25, 1994, p. 13A.
5. To contact the American Society for Quality Control (ASQC) or to inquire about the Koalaty Kid Alliance, write to ASQC, 611 East Wisconsin Ave., P.O. Box 3005, Milwaukee, WI 53201–3005.
6. To contact the American Association of School Administrators, write to AASA, 1801 N. Moore St., Arlington, VA 22209–1888.
7. The Pilot Education Criteria for the Malcolm Baldrige National Quality Award were released for comment December 16, 1994. These can be obtained by writing to the National Institute of Standards and Technology, United States Department of Commerce, Gaithersburg, MD 20899–0001.

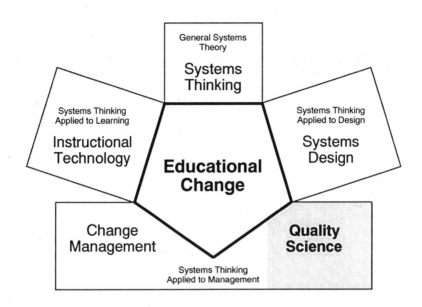

CHAPTER 7
QUALITY SCIENCE TOOLS

Quality science provides the set of reliable and proven methods for achieving excellence in an organization. Therefore, it is an important, even vital part in any improvement to the education enterprise. The previous chapter presented some of the basic concepts involved in quality science—key ideas that are part of our third technology. I described how work consists of processes, how quality science helps us control costs, how we can use data and information about processes to improve them, how to create virtual schools and virtual products, how to measure and increase customer satisfaction, and how to plan for corrective action. These ideas and actions have been highly successful in improving business, manufacturing, and other enterprises, and they can be just as successful in schools. This chapter presents information that will help you implement quality science ideas and approaches in educational organizations.

Quality Progress (the most widely read publication among people who are responsible for quality improvement efforts in their organizations) recently published a series of articles on what are commonly called the "Seven Tools of Quality." These seven tools are basic to quality science. By using these seven basic tools, people can successfully establish quality improvement processes in an educational system.

> **BEFORE YOU TRY . . .**
> - Before you try to solve a problem, define it.
> - Before you try to control a process, understand it.
> - Before trying to control everything, find out what is important.
> - Start by picturing the process.
>   —*John T. Burr*, **Quality Progress** *(June 1990), p. 64.*

Most books on quality contain a chapter or two about these tools. However, very few offer examples from education. Although many of the processes and activities that are involved in running a school system (such as clerical tasks and management functions) are similar to those that occur in any large organization, it is helpful to see examples specifically related to the education enterprise. In this chapter, all of the examples are drawn from education, and the language describing the application of these tools is tailored to educators.

In this chapter, I don't provide a lengthy discussion about the statistics that go along with the tools. I just describe the type of data that is collected for each and how to use that data to make effective system improvements. While it is true that quality science requires extensive use of data and knowledge of some basic statistical concepts, it does not require a well developed knowledge of advanced statistics. Any person with a basic knowledge of math and the ability to read charts and graphs can use the tools described here. Essentially, these basic tools are merely ways to organize and visually display data. In most cases, staff, teachers, or students can collect the data themselves, and they are happy to do so because it gives them more responsibility and control over their work. By using these tools, administrators, teachers, and students can come to understand the processes that they use in their work so that they can control the processes and improve them. Without tools to help define problems, we will forever be "putting out fires" rather than improving the system.

THE FIRST TOOL:
CAUSE AND EFFECT DIAGRAMS

Recall from the previous chapter that one of the critical steps in quality improvement is to identify the sources of variation in system processes, so that we can distinguish between random and special cause variation. The goal is to seek out special cause variation in the process. One of the easiest ways to determine the sources of variation is the cause and effect diagram. Cause and effect diagrams (also known as fishbone diagrams because they

Quality Science Tools

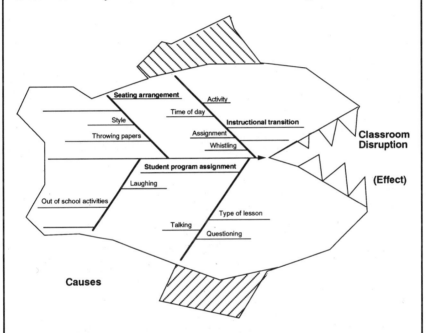

REDUCING LEARNING DISRUPTIONS

A team of four sixth grade teachers, a counselor, and the principal of LeRoy Martin Middle School of the Wake County Public School System in Raleigh, North Carolina used a cause and effect (fishbone) diagram to identify the causes of classroom disruptions in their school. The causes that they identified are listed along the "bones" of the fish. All of these causes contribute to the "classroom disruption" effect. Note that some of the causes are random variation (out of school activities, time of day, etc.). Others (shown in bold type) are special cause. These are the ones that the teachers chose to work on as the most effective way to eliminate or reduce classroom disruptions.

—Adapted from David L. Bayless, G. A. Massaro, E. Bailey, D. Coely, R. Holladay, and D. McDonald, "The Quality Improvement Management Approach as Implemented in a Middle School," ***Journal of Personnel Evaluation in Education,*** 6, No. 2 (December 1992), pp. 191–209. Used with permission.

resemble the skeleton of a fish) are often used to examine the factors that may be causing variation. The problem statement appears at the head of the fish. The bones represent the major causes that could explain why the problem exists. (Three different charts are shown here.)

Usually, these diagrams are constructed in a group brainstorming session. Each cause is listed on one of the "bones" of the skeleton and more causes are added as group members identify them. Eventually, the group members come to the point where they feel they have identified all the major causes that contribute to the problem. Once all of the causes that influence a situation or problem are identified on the diagram, it becomes fairly easy for the group to identify those causes that they have some control over—the special causes. These can be highlighted in **bold** on the diagram.

The main "bones" on the fishbone diagram can be grouped into major categories such as student, faculty/staff, curriculum, and outside influences. This makes it easier to see the major areas that cause variation. Factors grouped under *curriculum* and *faculty/staff* will most likely be the sources of "special cause" variation. These are the ones that we can eliminate and solve.

The cause and effect diagram is invaluable for analyzing virtually any issue or problem requiring attention, and it can be easily learned by people at all levels of the organization (including students) and applied immedi-

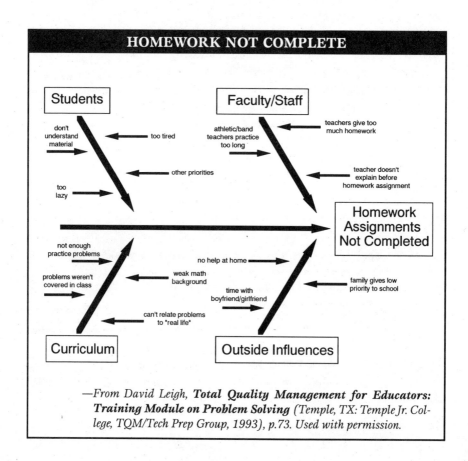

—From David Leigh, **Total Quality Management for Educators: Training Module on Problem Solving** *(Temple, TX: Temple Jr. College, TQM/Tech Prep Group, 1993), p.73. Used with permission.*

Quality Science Tools

BOBBY'S MATH SCORE

This fishbone chart was constructed by a fourth grader at St. Rita's School in Racine, Wisconsin. The chart shows the variables that could affect Bobby's scores in math. Some variables are under Bobby's control (effort, inclination to study, penmanship). Other could be controlled by mom or dad (limited distractions at home, some tools to aid study, availability for questions). Still others could be controlled by the teacher (clarity of questions, distractions at school, classroom environment, attitude toward learning).

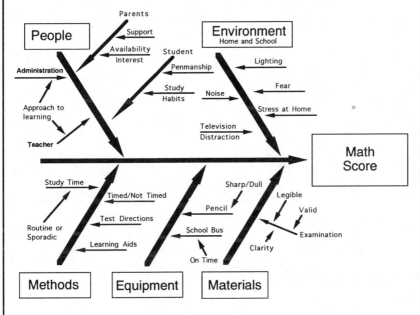

—From Mike and Jerry Stoecklein, "What Do Grades Mean?" ***Quality Goes to School: Readings on Quality Management in Education*** *(Arlington, VA: American Association of School Administrators, 1994), p. 213. Used with permission.*

ately. While a cause and effect diagram can be developed by an individual, it is best when used by a team. One of the most valuable attributes of this tool is that it provides an excellent means to facilitate a brainstorming session. It focuses the participants on the issue at hand and immediately allows them to visualize the different aspects of the problem. Again, we are interested in identifying those aspects of the problem that we have some control over, the special causes. We can then begin to work on these factors, thus improving the process and reducing the amount of special cause variation.

THE SECOND TOOL: CONTROL CHARTS

Education managers often base their decisions on data provided on a daily, weekly, or monthly basis. These data are usually displayed in reports with rows and columns of numbers. The problem with these types of reports is that they make no distinction between random and special cause variation. To find special cause variation, education managers must look at the data in a different way. Using a control chart is one way that makes it easy to distinguish random variation from special cause variation.

The control chart below shows daily absenteeism data over a period of 18 weeks. Similar charts could be used to show frequency of discipline referrals, classroom interruptions, scores on classroom tests or standardized tests, or any other type of information that can be tracked and plotted over time. The chart on the next page, for example, shows the amount of time spent on learning activities.

This presentation is different from typical numerical presentations that just list numbers on a sheet. The control chart has two important advantages. First, the numbers are plotted to show the pattern of occurrence over time. The second advantage has to do with the dotted lines along the top and bottom of the chart. These lines show the upper and lower "boundaries" of the data. Quality workers call the upper boundary the upper control limit (UCL) and the lower boundary the lower control limit (LCL). These boundaries are constructed according to statistical formulas from data collected on the process.[1]

How does such a display help managers? It shows them the special cause variation. Points that lie outside of the control limits or that fall into specific patterns are signals of special cause variation. In the absenteeism chart, there are two points outside of the control limits; a very low atten-

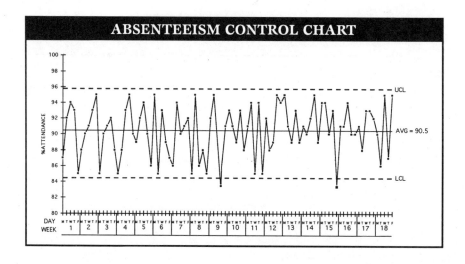

dance day occurred in the 9th and again in the 15th week. Thus, the manager can investigate what was unique about these particular instances, and will likely identify a problem that can be productively corrected. Other than these two instances, the variation in absenteeism is most likely due to random causes arising from ever-present, uncontrollable factors (illness, personal or family reasons). In the case of this graph, to ask why absenteeism was higher or lower on other days is a "low-yield" strategy, meaning that the manager won't get much payback for the time and resources expended in trying to answer the question. Worse, investing time in chasing random causes detracts from other, more productive tasks. Chasing random causes is called "tampering." Tampering is trying to counteract random causes. It is pointless because random causes cannot be controlled.

Managers can only make sound decisions if they understand random and special cause variation and know how to tell the difference. Otherwise, they are likely to spend most of their time tampering. Real improvements cannot be made until tampering stops. Control charts such as the one below showing learning activities data provide the quickest, surest way to determine the appropriate reaction to variation. (While there are no points shown here below the LCL, clearly there are major differences in amounts of time per day spent on learning activities, and further review may be warranted.)

Control charts show us what a system is capable of doing. They do not tell us whether the results are "good enough." The attendance rate, gradua-

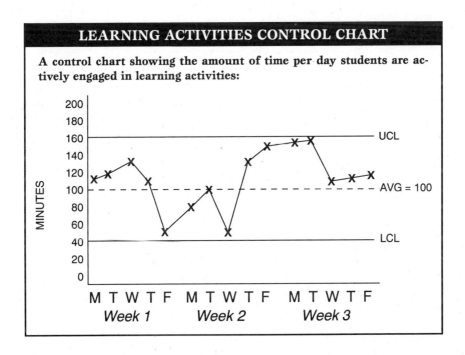

tion rate, or test results may not be what we hope to achieve. The control chart simply indicates what the system is capable of doing. The only way to improve a system that only has random variation is to fundamentally change it by instituting totally new procedures, materials, methods, or equipment.

THE THIRD TOOL: HISTOGRAMS

A picture can be worth a thousand numbers, especially when the picture is a histogram (sometimes called a frequency chart). A histogram is simply a bar type chart that summarizes a set of data. (See next page.)

When we use a histogram, we are quantifying some aspect of our process; we can then make management decisions based on facts, not opinions. Histograms often provide eye-opening, sometimes startling information in a format that is easier to explain to co-workers and managers.

Again, what we are looking for by using the histogram is variation, especially special cause variation. Displaying data in a histogram format makes patterns of variation much easier to see. Once we have a better understanding of the pattern of variation that exists in the process, we are much more likely to take actions that will result in positive, lasting improvements. We also know what we should investigate further. If we feel we need to, we can plan for direct observation of the process and collect additional data before we begin implementing remedies.

THE FOURTH TOOL: CHECK SHEETS

Quality improvement is an information intensive activity. Information helps us answer questions. Most organizations, including schools, have vast stores of data and facts about their operations. However, these data and facts may not be in a form that provides useful information for decision making. In many instances, it is the absence of clear, relevant information that allows a problem to continue unsolved for so long.

A check sheet is a simple data recording form that has been specially designed to enable the user to readily interpret results from the form itself. Instead of just recording observations as numbers and writing them as numbers in a column, the check sheet allows the observers to see what is happening in the process as they observe it. Notice the example here of a simple check sheet for recording the number of students using one of the computer labs in a school. Note that the form was designed to allow the observers to enter the number of students using the computers on a grid. As the data are entered, the observers will begin to see a pattern form. Because of this, the sheet provides more than a record of the data—it enables the simultaneous analysis of trends in the data.

HISTOGRAM: GUIDANCE APPOINTMENT SCHEDULING

This histogram shows data collected by the counseling staff at Springfield High School in Springfield, Pennsylvania. Senior students complained of long waits to see guidance counselors. Following the quality science approach to managing processes, the guidance counselors collected data and constructed a histogram to show the average number of days that it took for a students to get an appointment with a counselor.

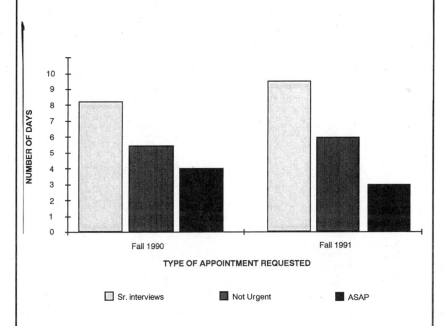

The data revealed that students had to wait about 8 days to see a guidance counselor unless the nature of the appointment was urgent, in which case it still took 4 to 5 days. The counseling staff developed a new process for scheduling appointments. Data collected the following year revealed that the new process had changed the situation only slightly. Urgent appointments were now being scheduling within 3–4 days but others were taking longer.

—From Roger Place, Tracy Kosman, and Katherine Vitale, "TQM: A Community College and School District Partnership," **Quality Goes to School: Readings on Quality Management in Education** (Arlington, VA: American Association of School Administrators, 1994), p. 198. Reprinted with permission.

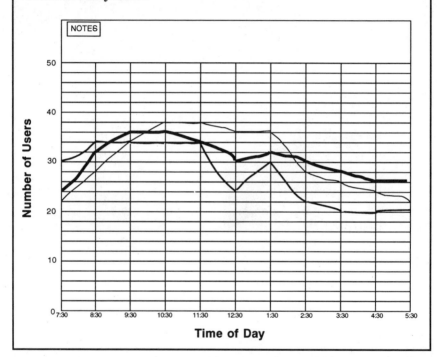

THE FIFTH TOOL: PARETO CHARTS

Named after the turn of the century economist Vilfredo Pareto, the Pareto Principle states that 80% of the problems come from 20% of the probable causes. What this means in most organizations is that we can obtain the most improvements by focusing our efforts on a few sources of problems. An organization may identify 50 or 100 problems, but most of them, according to the Pareto Principle, can be attributed to a few sources. If we can identify these few vital sources, we can solve most of our problems. The task of an effective manager is to focus improvement effort on the vital causes and not be distracted by the trivial many.

Quality Science Tools

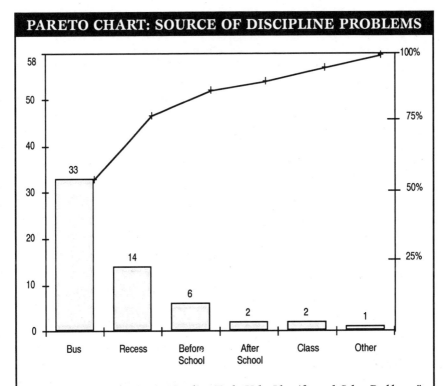

—From Peter Dedominici, "Quality Tools Help Identify and Solve Problems," ***Quality Goes to School: Readings on Quality Management in Education.*** *(Arlington, VA: American Association of School Administrators, 1994), pp. 185–187. Used with permission.*

The Pareto chart is a way of categorizing and prioritizing problems. Data can be collected on student absences, staff time usage, student time-on-task, school building vandalism, number of assignments that have to be redone, parent inquiries or complaints, number of errors or any other problem associated with learning or school. The data should be organized as shown in these Pareto chart examples. The most frequent cause is placed on the left, and the other causes are added in descending order of occurrence. The scale on the right side of the graph indicates the percentage of the total problem accounted for by each cause.

The first Pareto chart given here shows actual data collected by administrators at Denver Place School in Wilmington, Ohio on the sources of discipline cases at that school. The data would probably be similar at other schools. The Pareto chart makes it quite obvious which sources of problems need to be reduced or eliminated to have any real impact on the problem. In

PARETO CHART: OUT-OF-SCHOOL SUSPENSIONS

The following Pareto chart shows data collected by a team of administrators in Springfield, Pennsylvania on out-of-school suspensions. Claims were being made that their current out-of-school suspension policy was not effective. The team was interested in the number of repeat suspensions, that is, the number of times an individual students was suspended.

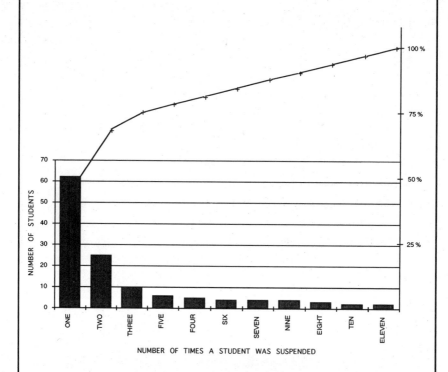

The Pareto chart revealed that 50% of suspensions were one-time only situations, with another 18% occurring twice in a year. Thus, the data did not bear out the claim of repeated suspensions; in fact, it showed out-of-school suspension to be a highly effective disciplinary tool for that district.

> —From Roger Place, Tracy Kosman, and Katherine Vitale, "TQM: A Community College and School District Partnership," **Quality Goes to School: Readings on Quality Management in Education** (Arlington, VA: American Association of School Administrators, 1994), p. 199. Reprinted with permission.

this case, the major source of discipline problems is on the school bus. This, then, became the focus of the Denver Place administration as they attempted to address the discipline problem. They decided to institute a peer discipline program called Fussbusters on the busses and in other areas of the school. As a result, they were able to reduce the number of discipline cases by 69% overall and by 73% on the school buses.

An organization that focuses on quality and efficiency should not allow anyone to begin work on a problem until he or she has developed a Pareto chart. The chart has to show that the problem being worked on is the most important one at the time. The alternative is to allow people to spend valuable time and energy solving trivial problems.

THE SIXTH TOOL: SCATTER DIAGRAMS

Sometimes two causes act together to create a problem. For example, the number of students who drop out of school may be related to the time of year. Plotting the number of students dropping out across several months may reveal the most critical time to take action.

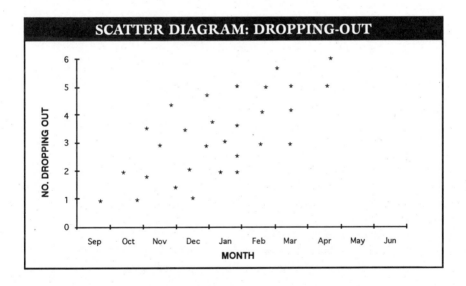

THE SEVENTH TOOL: FLOWCHARTS

The six tools discussed so far help us identify sources of special cause variation in our system. These are areas that we can directly influence and control to improve our processes. Once we have identified the major

GRADE REPORTING PROCESS

Middle School and High School Grade Reporting/Report Card Procedure:

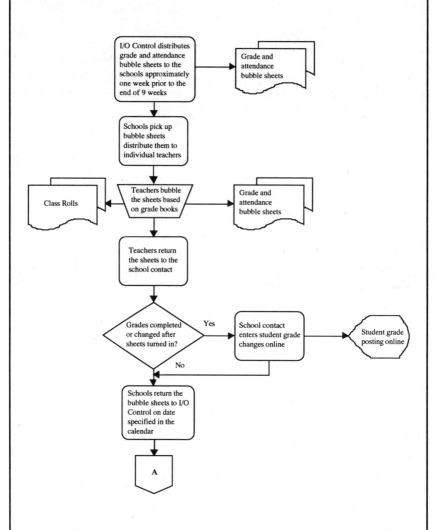

—From Florida State University, Center for Educational Technology, Quality Systems Work Group, Florida Schoolyear 2000 Initiative, **Quality Savings: Reducing the Cost of Poor Quality** (Spring, 1993).

ATTENDANCE MONITORING PROCESS

This flowchart shows the process for attendance monitoring used at a school:

- Bell rings
- Teacher takes class attendance
- Any students absent?
  - No → Check "No absence" on class roster
  - Yes → Check absent students' names on attendance roster
- Send roster to office
- Cause of absence known?
  - Yes → Complete absence record → File
  - No → Call student's home
    - Phone answered?
      - Yes → Ask why student not in school → Complete absence record
      - No → Line busy?
        - Yes → Call again → (back to Phone answered?)
        - No → Note: no contact

—From David Leigh, **Total Quality Management for Educators: Training Module on "Focus on Processes"** (Temple, TX: Temple Jr. College, TQM/Tech Prep Group, 1993), p. 16.

sources of special cause variation, the next step is to decide what to do about it. The first step in trying to improve a process is to construct a flowchart that shows the major steps of the process. This usually involves getting the people who use the process together and agreeing on the steps involved. Usually, in such discussion, disagreements arise among the team members as to what the process really is and how it actually works. Resolving these disagreements is valuable in and of itself, and arriving at an agreement about what the process is allows us to improve that process.

This insert on page 140 shows a process flowchart constructed by one school district. The chart depicts part of the process for reporting grades in that district. The chart was developed by a team of administrators and staff from the district office and schools. After describing the process being used, the district decided to make several changes in the process. In the case of this particular district, these changes resulted in a personnel hours savings of $2,600 per year and savings in computer supplies and computer time of $532 per year.

Flowcharting important processes in the organization is a critical step to quality improvement. The people who work in the process will understand it better. They can begin to control it instead of being victims of it. Once the process can be seen objectively in the flowchart, improvements can be identified. People realize how they fit into the process, and they can visualize their "suppliers" and "customers" as a part of that overall process. This leads directly to improved communication between departments, teams, and work areas. The people who participate in flowcharting work sessions become enthusiastic supporters of the entire quality effort. They will continue to provide suggestions to make even further improvements. Also, the flowchart becomes a valuable tool for training new staff or employees.

SUMMARY

The seven tools of quality have been illustrated in this chapter using examples from education—cause and effect diagrams, control charts, histograms, checksheets, Pareto charts, scatter diagrams, and flowcharts. These seven tools are basic to establishing a quality science process in education. The collection of the relevant information that is used by the tools generates interests, invites analysis, and highlights opportunities for improvement. When managers and other people in the organization systematically use these tools to define processes, determine problem causes, and monitor improvement, awareness is increased exponentially, and methods and processes can be improved to the point where they consistently provide good results.

FOR FURTHER READING

Howard Gitlow, Shelly Gitlow, Alan Oppenheim, and Rosa Oppenheim, *Tools and Methods for Improvement of Quality* (Homewood, IL: Irwin, 1989).

Julie E. Horine, "Reading, Writing, and Quality Tools," *Quality Progress* (October, 1992), pp. 33–38.

Kaoru Ishikawa, *Guide to Quality Control* (White Plains, NY: Quality Resources, 1982).

Roger Kaufman and Doug Zahn, *Quality Management Plus: The Continuous Improvement of Education* (Newbury Park, CA: Corwin, 1993).

David Leigh, *Total Quality Management for Educators: A Ten Part Series* (Temple, TX: Temple Jr. College, TQM/Tech Prep Group, 1993).

J. Stephen Sarazen (Ed.), "The Seven Tools of Quality," *Quality Progress* (June, 1990 to December, 1990).

Peter R. Scholtes, *The Team Handbook for Educators: How to Use Teams to Improve Quality* (Madison, WI: Joiner Associates Inc., 1994)

SOFTWARE

Some software tools are available to help organizations apply quality science and improve critical processes. These software tools are designed for businesses but are applicable to schools as well.

LearnerFirst Benchmarking 1.0. LearnerFirst, 1075 13th St. South, Birmingham, AL 35202. Tel. 205-934-9182. Available through ASQC, 1-800-248-1946. The innovative software application breaks down the complex benchmarking process into simple, step-by-step activities for successful benchmarking projects every time. This interactive software program will help you determine what to benchmark and how to do it, improve and streamline an existing benchmarking program, assess the strength and weakness of your organization's internal operations, increase the effectiveness and efficiency of your present practices, and provide a way to improve customer satisfaction.

LearnerFirst Process Management. LearnerFirst, 1075 13th St. South, Birmingham, AL 35202. Tel. 205-934-9182. Available through ASQC, 1-800-248-1946. This interactive, how-to software provides a road map to improving the performance of your organization's most critical processes. The programs help you apply process management concepts throughout your organization, and ensure high levels of customer satisfaction and loyalty. Learn how to become a fast, flexible, high-performance organization.

CATALOGS

The American Society for Quality Control publishes a catalog of helpful books, software, video, and audio resources. The catalog offers resources on customer satisfaction, service quality, education, the human side of quality, standards, statistics, teamwork, and more. To obtain a catalog, call ASQC at 1-800-248-1946.

NOTE

1. In statistical terms, the UCL and LCL are set anywhere between one and three standard deviations above and below the mean, depending on historical trends in the data. Some guidance from someone with statistical knowledge is essential in constructing the control limits.

PART IV

The Fourth Technology: Change Management

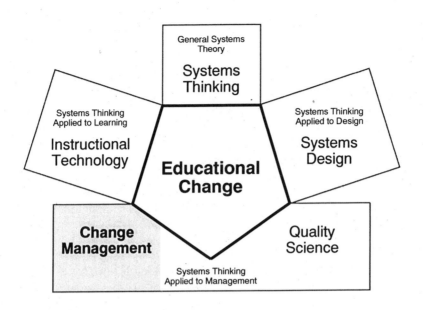

CHAPTER 8

CHANGE MANAGEMENT

"Rapid change is the new status quo."
—**Daniel Burrus**

Systems thinking, systems design, and quality science (the three technologies you have read about so far) are all necessary to create an effective, efficient, and people-oriented education enterprise. Creating a new education system that can meet increasing demands will require radical change. It will involve reconceptualization and discontinuity from the initial system—a revolutionary jump.

If we are to create virtual schools and virtual products that meet people's needs, we will have to create a new kind of enterprise based on choices, sound instructional methods, cost-effective use of new electronic tools, high quality instructional materials, and the ability to deliver these anytime, anywhere. This new enterprise will allow a greater number of students to master reading, writing, mathematics, history, geography, science, and the other basic subjects at an encouraged pace and to apply them together to solve meaningful problems. They will do it, not in the traditional way (with teachers lecturing to groups of students), but through high tech learning resources and effective strategies for developing, reinforcing, and assessing basic and advanced skills. Education leaders who wish to guide their organizations into these new realms will need to understand the dynamics of change and be able to draw upon the necessary skills for managing that change. Change management is the key technology that education leaders must use to guide change successfully.

CHANGE MANAGEMENT

Some people may think it strange to talk about *managing* change. Most people tend to think of change as chaotic or random rather than as a predictable process that can be managed. However, the recent experiences of businesses and other organizations that have undergone major transformation have shown us that change is not so chaotic or random. Successful implementation of change in an organization follows distinct patterns—there are identifiable forces that shape this success. Once we understand these patterns and forces, it becomes possible to deal with them in a planned way rather than haphazardly as they surface one at a time.

PERSPECTIVES ON CHANGE

"If organizational change is to be anything more than an interesting topic of conversation, the connection between the human being, the change itself, and the organization must be understood."—Frank Mellon, Team Leader, Dow Corning Corporation

"The key executive skill in the 90s will be managing change . . ."—Michael O. Bice, President, Allegheny Health System

"Slowing down the rate of change in our fast-moving world is simply not an option. Increasing our ability as individuals and as organizations to assimilate more and more change faster and stay effective is the answer."—Donald M. Arnoudse, Vice-President CSC Index, Inc.

"Having a good idea is only half the story—the more important half is getting the idea implemented and reinforcing its use throughout an organization."—William C. Byham, President and CEO, Development Dimensions International and author of ***Zapp! The Lightning of Empowerment.***

"Learning to manage change proactively is probably one of the most fundamental skills a manager must have to position himself and his organization on the scale that goes from survival to market leadership."—Maurice Cazes, Executive Development Consultant, IBM Brazil.

"Success—even survival—today requires ongoing change. The challenge of a leader is to be a positive instrument of change."—David A. Cole, Chairman, Kurt Salmon Associates, Inc.

Managers and management scientists, organizational psychologists, and others have learned a great deal as they have experienced or observed the process of transformation and change in organizations over the last few decades. As a result, we now know many things that managers and leaders can do to skillfully bring about desired changes in organizations. What we have learned is that knowing *what* to change is only half the story—the even more difficult part is *getting the change implemented and routinized throughout the organization.*

Managing change is a process that can be and has been studied. Learning to manage change proactively is quickly becoming *the* fundamental skill for successful managers and leaders. Without quality leadership and skillful management, even the best ideas are never implemented. Without good management and ongoing support from their leaders, those lower in the organization become disillusioned and, in time, cease to continue the change effort.

ORIGINS OF CHANGE MANAGEMENT

The source of this new knowledge about change comes from the business management literature on organizational development and organizational transformation. This is a relatively new field of study for management scientists and scholars. It arose during the 1970s and '80s as managers and organization leaders were forced, because of increased competition, to oversee drastic reshuffling of every aspect of their organization—its mission, goals, structure, and culture. Observers of these changes began to call this "second-order change" (to distinguish it from first-order changes, which are only minor improvements and adjustments in the system). So this field deals with the study of radical, basic, total change in an organization.

Change management is on the cutting edge of management science. Today, it is in the forefront of the management literature. It draws upon the insights, observations, and ideas from pioneering, innovating thinkers around the world. This chapter can, in no way, provide a comprehensive survey of the practice and theory involved in understanding and managing change. What the chapter does present are the major ideas that successful leaders need to understand and key strategies that can help them succeed in accomplishing the goal of change.

CHANGE MANAGEMENT AS A TECHNOLOGY

Some may not agree with me that change management is a technology. They are not wrong. Neither am I. There are surely differences of opinion about what constitutes a technology (see Chapter 1). There is no question, however, that there now exist specific, reliable, and known strategies, steps, and principles for helping leaders succeed at initiating and implementing large scale change.

Change management, then, is a set of actions and tactics for understanding and guiding change in an organization. These strategies and tactics have been shown to create a more positive climate for change that allows leaders and the people working in the organization to successfully implement change. These strategies and actions also allow leaders to deal with the human side of change in a proactive way by understanding the personal and emotional responses and circumstances that invariably arise during the change process.

Change management is a technology for guiding change that allows leaders to:

- help people prepare for, accept, and become committed to the change;
- understand the roles of different people in the change process;
- know how much change is too much;
- understand the stages of change and the proper management strategies for each stage;
- know when to expect resistance; and
- know how to turn resistance into something constructive instead of destructive.

CHANGE IN EDUCATION

Nearly all people who work in education have come to realize that changing schools is complex. Every time a new initiative is launched, there are side-effects and unforeseen consequences that force even the most optimistic to accept the systemic nature and mind-boggling complexity of attempting change in education. The complexity and inter-connectedness of all that occurs around change can make the idea of changing schools frightening and overwhelming.[1]

Educators work in an environment in which they have been literally bombarded by one announced change followed by another. However, they have seen very few of these changes ever materialize at their own level of activity. Educators who have heard school principals, district, or state level education officials announce changes, yet have not seen results, become disillusioned. They learn to ignore these announced changes and to question any new change by asking themselves, "Is this just the newest thing to come down the pike or are they serious this time?" or "Maybe if we just wait awhile, this will go away like all the others."

STAGES OF CHANGE

Earlier, I mentioned that management experts have learned about change management from observing the major transformations that have taken place in many organizations over the past two decades. One of the major observers of this transformation process has been Daryl Conner. For nearly 20 years, Daryl Conner's group, Organizational Development Resources Inc. (ODR), has studied how major change takes place in large organizations. They have observed that change nearly always follows a series of predictable patterns. One of these patterns consists of three distinct stages: *preparation, acceptance,* and *commitment.*[2] For change to be successful, leaders must help those they supervise to get through all three phases. In the preparation phase, those who will be involved in the change effort need to become aware of the need for change and understand that the proposed changes will affect their personal roles and responsibilities. This does not necessarily mean that everyone involved will understand the specifics of exactly how the proposed changes will affect them. They may be confused or unclear about the exact nature of the change. During this phase, they may even disagree that the changes are necessary. The role of the leader is to provide information that will help to lessen the confusion, supply information about why the changes are necessary, mold the change

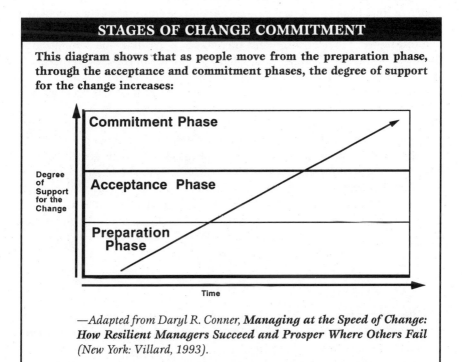

STAGES OF CHANGE COMMITMENT

This diagram shows that as people move from the preparation phase, through the acceptance and commitment phases, the degree of support for the change increases:

—Adapted from Daryl R. Conner, ***Managing at the Speed of Change: How Resilient Managers Succeed and Prosper Where Others Fail*** *(New York: Villard, 1993).*

based on constructive input from others, and ultimately, help people in the organization be ready to move into the next stage—acceptance.

People enter into the acceptance phase once they understand the nature and intent of the change and begin to accept the idea that the change is necessary. The indicator that people have moved into the acceptance phase is that they now view the change as a reality. This does not mean that everyone's attitude about the change is positive. Even after the change has been molded by constructive input, there will still be both positive and negative reactions. But the change is now seen as primarily positive, in spite of some of the negative perceptions that may remain. Ultimately, for the change to be successful, the organization leader must assure that key people within the organization make it through the acceptance stage. Reluctant people can stop a change effort in any number of ways (slowness of action, negative word-of-mouth among co-workers, etc.).

The final stage is the commitment stage. This phase doesn't begin until people (both from a personal and group standpoint) decide to embrace the change. Even though they have decided to go ahead with the change, people will still have reservations about certain aspects of the change (start-up issues, details of implementation, etc.). The role of the manager at this stage is to maintain an open work environment by encouraging open discussion to identify and solve problems and build team spirit and group commitment. The commitment stage may represent a decision to adopt the change for a short- or long-term test. The result of the commitment stage may or may not be adoption, because the results of the test may not prove favorable. If, after the test period, we determine that the change is positive, the commitment stage leads to institutionalization, where people no longer view the change as tentative. It has now become a permanent aspect of the organizational structure or operating procedures. If the commitment stage has been carried out successfully, the people within the organization will view the change as "their own." They will have become personal advocates for the change and will continue to use the new procedures with enthusiasm and commitment. On the other hand, if the commitment stage has been circumvented, people may have adopted the change due to fear of reprisal or due to the presence of severe consequences for not conforming. If this is the case, the change may still fail due to half-hearted efforts and lack of commitment. Also, if those leading the change leave their positions, things may return quickly to the way they were prior to the change.

KEY ROLES

Another important observation made by ODR is that successful change is dependent on the successful performance of key people in four distinct roles: These roles are sponsors, agents, advocates, and targets.

Sponsors

Sponsors are people in positions of authority—individuals who have the power to authorize a change, to set new priorities, to reallocate budgets, and to provide the time people will need away from their regular jobs to design the change. Sponsors communicate the new priorities to the people in the organization and provide the proper reinforcement, budgets, and time to ensure success. Sponsors are also responsible for creating an environment that enables these changes to be successfully implemented and institutionalized.

If a sponsor is to be successful, he or she must invest resources, devote some of his or her own personal time in publicly endorsing the change, and stick to the goal over the long haul. One of the sponsor's roles is to keep the change effort on track by rejecting solutions that promise short-term benefits but which may not prove fruitful in the long term.

Agents

Agents are the individuals in the organization who are responsible for making the change happen. Their job is to help put the change effort in place—to get it going and keep it going.

Agents develop a plan to accomplish the change, diagnose potential problems that may limit implementing the change, and carry out the steps necessary to execute the change effectively. The participation of change agents who possess these skills is a crucial factor in the success of any change project.

Advocates

Advocates are individuals who want to make a change but who lack the power to authorize it. Advocates generate ideas for change, but these ideas die if the advocates fail to gain support from the sponsors who can approve and authorize the ideas. Anyone can be an advocate, but advocacy (wanting a change to occur) is not enough. The critical ingredient for successful change is sponsorship from those who have authority to make it happen.

KEY ROLES IN THE CHANGE PROCESS
SPONSORS
AGENTS
ADVOCATES
TARGETS

Targets

Targets are the individuals who must actually change. The term *target* is used because these people are the focus of the change effort and play a crucial role in the short-term and long-term success of the change. To increase the likelihood of success, targets must be well informed. They must understand the changes they are expected to accommodate and the reasons for those changes. They must help plan the implementation process.

SUMMARY OF KEY ROLES

Sponsors, advocates, agents, and targets represent four distinct roles that people perform during the change process. Each role requires a specific set of skills and abilities. For example, sponsors must demonstrate commitment and know how to communicate that commitment to others within the organization. Good sponsors must be willing to apply meaningful incentives and pressure to produce the desired result. Advocates must possess effective communication skills so that they can convince others of the desirability or necessity of the change. Change agents must possess good planning and project management skills. They must understand the stages of changes and plan appropriate steps to guide people and the organization through the stages. Targets must comprehend their role also. They should be able to participate in the process, be given opportunities to provide input, and receive appropriate support, preparation, and training.

Each person in the organization must understand the nature of the role he or she is performing and what his or her job is. The same person may, at different times, play the role of sponsor, agent, advocate, or target. For example, John, a principal in an elementary school, may be an advocate for a particular change. He may also be the sponsor of the change in his school. Along with others who work with him, he may play the role of change agent. Finally, the change may require that the principal himself change, so he is a target of the change as well. In this example, the important thing for John to remember is that these roles require different skills and abilities. He must understand that his role as sponsor is different from his role as change agent, and he must be able to utilize the skills and abilities appropriate for functioning in those distinct roles.

STAGES OF CONCERN

Change, by its very nature, is disruptive and uncomfortable. People's routines are affected; their plans and expectations altered. Because of the disruptive and discomforting nature of change, it is natural that people will

Change Management

experience a variety of concerns, emotions, and reactions as they experience the process.

Gene Hall, co-author author of the book *Change in Schools*,[3] has studied the concerns, feeling, and reactions that people have to change as they are involved in it. What Hall discovered was that people's feelings and perceptions about a change can be categorized into three types of concerns:

- concerns about how the change will affect them personally (self concerns);
- concerns about "how to do it" (task concerns); and
- concerns about how to make the change better (impact concerns).

Also, Hall has documented that, in a well managed change process, people's concerns progress through these three stages in a predictable pattern. The three stages of concerns correspond closely to the three stages of change (preparation, acceptance, and commitment) observed by Daryl Conner.

Self Concerns

When a change effort is in its early stages, people are very likely to have concerns that relate to their own understanding about the change and their personal involvement in it. Typical questions are:

- "What is the change and how is it similar to or different from what I am already doing?"
- "What is the reason for the change?"
- "Who is endorsing it and why?"
- "How is it supposed to work?"
- "When will the change begin?"
- "Will I be able to do it successfully?"

All of these are *self-concerns* because they relate to the individual's personal role and understanding. Personal concerns will likely be intense during the initial phases of the change effort. Hall points out that it is common, during this stage, for people to be concerned about their ability to execute the new program as expected and about making mistakes that might make them look foolish. If they are resistant, they may also try to characterize the change as nothing new, but as something they have always done, or used to do, or that they already tried. These types of expressions may be used to convince themselves that they really do not have to change.

The correct approach for dealing with these types of concerns is to realize that they are normal and expected. The effective leader will provide full and timely information about the change, about why the change is different from what is being done or from what has been tried before, and about why the change is necessary at this time. The leader will allow, even solicit, open discussion and voicing of concerns, and will devote a good deal of personal time to establishing rapport and increasing interpersonal communication and support during this stage.

Task Concerns

The second category of concerns is *task concerns*. Task concerns typically become prominent as implementation begins. Task concerns are concerns about how to do it.

- "How do we get the time to do this?"
- "How do we arrange to have the people and materials in the right place at the right time?"
- "How do we learn to operate the new computers?"
- "How will we schedule students?"
- "How will we communicate with each other?"

The useful approaches for dealing with task-related concerns include providing additional training, printed job-aids, mentors, technical support, question and answer sessions, or on-the-job help. The effective leader will realize and expect that people will not be able to perform new tasks or function in new roles without a great deal of support. The fact that *task* focused concerns are being expressed is a positive sign that the change has moved forward from the preparation phase.

Impact Concerns

Once a change is installed and operational, people may begin to have concerns about the effects of the change on students or parents and about how to improve the program or process.

- "I am concerned about how this change will affect the results we have been getting."
- "I have some ideas about something that would work even better."

During this stage, teachers, and others will have concerns about the effects of the change on students and what can be done to improve the effectiveness of the program. Once teachers become used to the new way of doing things, they may become concerned about finding even better ways to improve it. At this point, it is important for the leader to capture these concerns and ideas for further improvement and channel the energy of individuals into productive paths.

THE CONCERNS PATTERN

Again, Hall and his associates found that concerns tend to progress through these phases in a distinct and predictable pattern. *Self*-concerns will be most intense early in the change process and then will decrease as the process moves toward implementation. At that point, *task* concerns become prominent. Only after the change is in place and operational will *impact* concerns intensify. The pattern and intensity of concerns that the people in an organization experience is directly related to the speed of change and the magnitude of the change, as well as the degree to which the change involves modifying individual practices and behavior. All of these stages are normal, predictable, and even necessary. Too many changes happening at once, however, can bring on debilitating stress and dysfunction.

PROCEDURES FOR ASSESSING CONCERNS

Hall has identified methods and procedures for assessing people's concerns during the change process. These methods involve a combination of:

- informal face-to-face conversations (also called one-legged interviews);
- having people write out descriptions of their concerns; and
- use of open-ended concerns questionnaires.

Informal conversations or one-legged interviews (see the following insert) are especially powerful precisely because they are unobtrusive. These may take place in the hallway, in the faculty lounge, or in other ways. The effective leader will take advantage of these opportunities to ask sincere questions of people about their thoughts and feelings related to the change. The leader will also use formal group meetings as opportunities to solicit and *listen* to people's expressions of concern, looking for indicators and clues that suggest which stage(s) is being expressed. The leader should begin by asking open-ended questions (e.g., Tell me more about what you mean when you say you are concerned about _____). Such questions, which encourage open expression, will help clarify what the person's or group's concerns are. After a thorough airing of concerns, the leader should make an appropriate response, such as providing information or making specific recommendations.

If the change involves a large number of people, it will be helpful to collect concerns by having people respond to an open-ended question on a simple form (see the insert on page 159). A question such as "When you think about _____, what are your concerns?" is written at the top of the page. The name of the change, the project, or the innovation is written in the blank space and the remainder of the page is blank. This

> ### QUESTIONS TO ASK IN ONE-LEGGED INTERVIEWS
>
> (Use one or more, not all at the same time.)
>
> 1. How are you feeling about the change?
> 2. What is your reaction to the _____ aspect of the change? (Ask specifically about different aspects of the change.)
> 3. What problems or concerns do you have about the change?
> 4. How does the change affect you? Your job?
> 5. Do you have any reservations about the change?
> 6. What other suggestions do you have for this effort?
>
> Responses to these questions may cue additional questions.
> This method requires that the questioner be a good listener; allow respondents time to say all they wish without interruption; does not try to direct the responses or to put words in respondents' mouths; and give respondents evidence that he or she really is listening by asking elaborative questions, restating certain statements, and listening to the whole response.
> —Adapted from S. M. Hord, W. L. Rutherford, L. Huling-Austin, and G. E. Hall, **Taking Charge of Change** (Alexandria, VA: Association for Supervision and Curriculum Development, undated).

format provides people an opportunity to write one or more sentences describing their concerns about the change. The responses can then be analyzed to determine the stage of concern being expressed (*self, task,* or *impact*). The leader can present the concerns that were collected using this method back to the group in a group meeting, or if that is not possible, in a handout or newsletter. Again, after presenting the concerns, the should solicit proposals or recommendations for addressing the concerns. Presenting the concerns to those who submitted them and discussing them in this way has two benefits. First, allowing those who submitted concerns to see that the concerns have been read, analyzed, and compiled sends the message that management *is* listening. Second, it allows individual people to see that they are not the only one with a particular concern. This can bring about healthy discussion that will lead to positive solutions or resolutions to the concerns expressed.[4]

One additional note about analyzing concerns: Each sentence in the response should be analyzed separately and then the whole response considered collectively. For example, one response might be: *"(1) Almost every night I wonder if I'll be able to locate and organize the new materials I will be using the next day. (2) I can't yet prevent surprises that cause a lot of wasted time. (3) I wonder how my students will be evaluated."*

Sentence 1 indicates the person's concern about materials and their organization, which is a task concern. Sentence 2 is a concern about wasted

---

**USING OPEN ENDED QUESTION
TO ASSESS CONCERNS**

Date:_____

Group: _____

Please use this opportunity to express your reactions, concerns, or reservations to the Schoolyear 2000 Initiative.

When you think about the Florida Schoolyear 2000 Initiative, what are your concerns? (Please be frank and use complete sentences.)

_____
_____
_____
_____

—*This is an example of a concerns sheet that we used as part of The Florida Schoolyear 2000 Initiative, a large scale change project in Florida public schools. Note that respondents are encouraged to answer in complete sentences so as to provide enough information for accurate analysis.*

---

time, another task concern. Sentence 3 also reflects the need for information in relation to an impact concern. Collectively, these sentences show that this person's most immediate concerns are at the task level but the person is beginning to ponder impacts.[5]

Other points about responding to people's concerns:

- People tend to express only those concerns that are most important to them. Many will also have other concerns at the other stages, but they may not express them. The leader should realize this and not ignore the concerns that may exist at other stages.
- Concerns change over time, so they should reassessed periodically as the change effort progresses.
- There is nothing inherently good or bad about concerns at any particular stage. The important thing to realize is that the must take one kind of action to address concerns at one stage and a different action to help address concerns at another stage. Individuals who have high personal concerns will have little or no receptivity to assistance that is directed toward task oriented concerns or impact concerns. They must respond appropriately first to their personal concerns. For example, arranging training for teachers who have task concerns would be effective. However, for teachers with intense personal concerns, the training would be of little or no value; in fact, it may serve

to intensify these persons' level of personal concerns. Once personal concerns have been reduced, task concerns can be addressed more effectively.
- Movement through the stages of concern cannot be forced, but, with appropriate support and assistance, it can be aided. At the same time, a lack of assistance or the wrong kind of support can interfere with progress through the stages.

JUST SAY NO!

All organizations have a limited capacity to assimilate change. Because of this, smart managers and leaders will resist the temptation to adopt every popular quick fix and instead keep the pressure on everyone in the organization to stay the course on the most important change effort. The leader has to realize that the popular quick fixes are merely distractions that will prevent or postpone arriving at the more enduring solution. Many principals and superintendents have trouble saying no to good ideas. One cannot conduct a major change effort in education by adding it to an already full plate of smaller initiatives. One must be willing to abandon some existing commitments. Often, concurrent programs and initiatives will have overwhelming and even contradictory requirements. The result is that the whole thing overwhelms the personnel in the organization and discourages them from supporting *any* new effort. Schools, like other organizations, have a limited capacity to absorb change. Trying to make too many changes at once may lead to burn-out, confusion, resistance from parents, resignations by employees, covert efforts to undermine the change, and other dysfunctional behaviors.

Because systems design requires investment of time and resources, education must enable teachers to find the time to do it by taking away some of the things they are currently doing with their time. You cannot continuously add more. Effective change managers must find ways to enable teachers to design and begin implementing system changes while the school continues to function. Many schools today are experiencing problems simply because they are attempting too much change at once.

SUCCESSFUL IMPLEMENTATION

Because change is a process and not an event, it has to be orchestrated carefully. Just because we have a great idea or solution does not mean that it will be implemented as intended. No matter how sound the proposed change may be from a logical or theoretical standpoint, successful implementation requires skillful management. The successful leader will plan a careful and detailed implementation strategy and will allow sufficient time and resources to support the implementation process.

Successful leaders view the implementation process as a continuum, with periods of progress intermixed with periods of delay and unforeseen obstacles. The leader is not surprised by these delays because he or she realizes that they are an expected part of the process. In other words, the manager of change has learned to expect the unexpected. Some call these delays or problems "UNK, UNKs," an abbreviation that stands for the *unk*nown *unk*nowns that we know we will encounter during the implementation process.

In addition to expecting the unexpected, the successful leader will teach those others in the organization to do likewise. A leader who sells "snake oil" by telling others that the change will be easy and painless sets up unrealistic expectations and, by this action, contributes to the failure of the change. When the real costs and unforeseen effects of the change begin to become apparent—when the "going gets rough," so to speak—people in the organization will retreat to the status quo. If all are expecting the road to be difficult and challenging, they will be much more likely to follow the implementation process through to the end.

RESISTANCE: A NATURAL RESPONSE TO CHANGE

Change is disruptive. It causes unforeseen effects. It creates stress. It makes people's emotions to rise. It is always uncomfortable. Because change is uncomfortable, it will always bring about resistance. This is true irrespective of whether the change is bad or good. Resistance may surface because the change is actually a bad idea. But even if the change is likely, from a logical or theoretical standpoint, to be very positive, there will still be some resistance. This is true simply because resistance is the natural human response to change.

Effective managers of change realize this and will not view resistance as a problem, or even worse, view those persons who are resisting as disloyal, ignorant, or close-minded. Instead, the effective manager of change will use resistance and resisters as part of the solution rather than seeing them as part of the problem. The key to making resistance part of the solution is to make resistance acceptable by allowing people to resist *openly*. Make it known to all, by both words and actions, that their suggestions and expressions of concern are encouraged and will be seriously considered. Effective leaders provide opportunities for resisters to air their suggestions or concerns openly and thereby allow appropriate action to be taken to address legitimate fears and concerns. If the change is actually a bad idea, this open discussion process will allow clearer heads to prevail and hopefully avoid a potential disaster. On the other hand, if the change is generally positive, open participation will likely make the change better. Legitimate problems can be identified and corrected, making the final result much more desirable and effective. Also, because more people have participated in the decision process, they likely have come to view the change positively, and will therefore be more inclined to assure that the change becomes institutionalized and that it will stick even after the other key personnel are gone.

Experience has taught us is that no one person or group can see the total effect of the change or how it will impact every aspect of the operation or every person in the organization. This is because the change is always embedded in a larger system. From the systems thinking point of view (see earlier chapters), any given change will have an effect on every other part of the system. The viewpoints, experience, and input from managers, parents, teachers, and others will all be different as they view the change as a part of their whole life's concern. Each individual or group that participates in the process can influence it for the better.

THE IMPACT OF "CORPORATE CULTURE"

Every organization, including schools, has a "culture" that pervades the entity. This culture consists of the beliefs and assumptions about how the organization runs and what its mission is. In the business world, this set of beliefs and assumptions about the business and its practices is called "corporate culture." Corporate culture is the way the business thinks about itself and what it does. Corporate culture is what makes up the business's image of itself. It is the habit patterns that have come to be expected in the organization. Every institution, every business, every corporation, every school, and every school district has a "corporate culture"—a set of habit patterns that affect how the business, corporation, or school functions.

The important thing to know about this culture is that it exerts great power over what people perceive as permissible and appropriate. The beliefs, values, and habit patterns that are embedded in the culture carry tremendous voltage. Like electricity, the culture sends its current throughout the entire organization and influences everything that occurs.

The culture has a life of its own. It has its own power and influence; its habits have grown strong, perhaps over a long period of time. It shapes the behavior of every person in the organization. It can be very controlling.

The organizational culture impacts change by driving away any initiative, decision, or action that is inconsistent with the existing culture. If the change we want to introduce is generally acceptable to the existing culture, then we can expect a high probability of success. If, on the other hand, the proposed change has very little in common with the current culture, or runs counter to current cultural trends, our chances of successfully achieving the change are slim. Changes that are incompatible with the existing corporate culture will likely fail.

The corporate culture that exists in our public schools system is strongly embedded in historical policy and tradition. Change tends to come

> **COMPONENTS OF THE CULTURE**
>
> Components of the Culture
> Rites and Rituals
> Accepted Paradigms
> Norms/Values
> Structure
> Defined Career Paths
> Leadership
> Reward System
>
> —Price Pritchett, *Culture Shift: The Employee Handbook for Changing Corporate Culture* (Dallas: Pritchett Publishing Company, 1993).

slowly because of the bureaucratic structure inherent in the system. The implications of such a culture are that educators must do one of two things: They can restrict themselves to introducing only those changes that are relatively consistent with the existing beliefs, values, and assumptions, or they can develop and resolve to build a new set of values and assumptions for teachers and colleagues and for the community that are more compatible with the necessary changes.

It is very difficult to implement a change that is not compatible with the existing culture. Because the existing culture was developed over a period of years, changing it is very, very difficult. In most cases, what you have to do is modify the change to be more in line with the culture, while at the same time taking steps to begin changing the culture. These steps can include downsizing, relocating or merging groups or teams, changing the reward system, promoting the vision, allowing more choices to students and parents, becoming more customer oriented, simplifying, and involving everyone. By doing these things, the effective change agent can begin to mold the existing culture into one that can accommodate the change—one that is driven by new ideas or by higher standards of effectiveness.

OFTEN-ASKED QUESTIONS ABOUT MANAGING CHANGE

Question. *Is change an event or a process?*

Answer. Change is a process that involves stages, steps, and phases. Those who think of change as an event have the false expectation that a one-time announcement, a policy mandate, a new curriculum guide, or the replacement of a principal will bring about change. Those who take this "event orientation" approach to change (see the insert below) will find that the changes they expected are never realized. The complexity and systemic nature of change have already been discussed. Whether you are trying to implement a single innovation or a large scale change, there is an evolution that has to be understood and a process that has to be followed.

Question. *What is the role of leadership in the change process?*
Answer. Leaders guide their organizations into new realms. They facilitate change from the status quo to a completely new operating framework. Change management is the set of skills the leader must use to be successful in initiating and implementing change. Leaders must manage change, but they are not just managers in the typical sense. Managers facilitate change only within the existing operating framework. Leaders facilitate change from the old framework to the new one.

EXAMPLES OF THE "EVENT ORIENTATION" TO CHANGE

1. School Improvement Teams follow a ten-step set of activities to develop a school improvement plan. The tenth step will be "implement." Note that the first nine steps were spent on developing the plan, but the planners anticipate that implementation will be tackled at the end by a single step.

2. Strategic planning processes that take one to two years of work by committees and teams to develop a 3–5 year vision and plan. Ultimately the plan is delivered to the rest of the organization with the assumption that it will be 'implemented.'

3. Curriculum development is done at the local or state levels. This process generally takes two to three years, with funding, extra resources, specialized personnel, and assigned time to create the new curriculum and curriculum frameworks. Then this is followed by a 'dissemination' of the curriculum/curriculum plan with the expectation that there will be 'adoption' by all schools, without a detailed plan for implementation.

4. Policy initiatives, policy mandates, or new laws are established by one or another governing body and then little or no attention is given to the necessary technical assistance, assignment of time, or release from prior rules or regulations, to facilitate implementation of the expected changes.

—Adapted from Gene E. Hall, "Characteristics of Change Facilitator Teams: Keys to Implementation Success," **Educational Research and Perspectives** 19, No. 1 (The University of Western Australia, June, 1992, p. 99).

Question. *Why is sponsorship so important?*
Answer. Sponsorship is important because it is the key element in initiating and sustaining change. Major educational change is not possible until it is has received endorsement, resources, and leadership from the top down. Unless the sponsor shows a wholehearted and ongoing commitment to the change effort (through public and private statements and through allocation of key budget resources) the project will eventually fade into obscurity and die. Other people in the organization can be charged with implementation, but only sponsors can legitimize the change.

Question. *Isn't change best implemented from the "bottom-up"?*
Answer. This question is based on the assumption that if the people at the bottom of the organization get all excited about and own an idea, it will then bubble up through the organization and result in successful change. However, this view is inconsistent with what we now know about roles in the change process.

Question. *What is the difference between small scale change and large scale change?*
Answer. The difference is not a matter of the amount of money being spent, the number of people involved, or whether the change affects the classroom, school, district, state, or all levels. Rather, the difference is in the degree to which the change deviates from people's expectation patterns. Small scale change is change that does not deviate much from the established patterns and direction of the organization. Large scale change significantly disrupts people's expectations. Small scale change is much easier to implement than large scale change but is less likely to have a major effect on the system overall.

FOR FURTHER READING

Change Management
Chris Argyris, *Knowledge for Action: A Guide to Overcoming Barriers to Organizational Change* (San Francisco, CA: Jossey-Bass, 1993).
Warren Bennis and Joan Goldsmith, *Learning to Lead: A Workbook on Becoming a Leader* (Reading, MA: Addison-Wesley, 1994).
Daryl R. Conner, *Managing at the Speed of Change: How Resilient Managers Succeed and Prosper Where Others Fail* (New York: Villard, 1993).
Patrick E. Connor and Linda K. Lake, *Managing Organizational Change* (New York: Praeger, 1988).
Thomas G. Cummings, Edward E. Lawler III, Gerald E. Ledford, Jr., Allan M. Mohrman Jr., and Susan Albers Mohrman, *Large-Scale Organizational Change* (San Francisco: Jossey-Bass, 1989).
Michael G. Fullan, *The New Meaning of Educational Change*, 2nd ed. (New York: Teachers College Press, 1991).

Michael G. Fullan, *Change Forces* (Bristol, PA: Falmer Press, 1993).
Michael Hammer and James Champy, *Reengineering the Corporation: A Manifesto for Business Revolution* (New York: HarperCollins, 1993).
Ronald G. Havelock, with Steve Zlotolow, *The Change Agent's Guide,* 2nd ed. (Englewood Cliffs, NJ: Educational Technology Publications, 1995).
K. Scott Hughes and Daryl R. Conner, *Managing Change in Higher Education: Preparing for the 21st Century* (Washington, DC: College and University Personnel Association, 1989).
Jerrold E. Kemp, *A School Changes* (Washington, DC: Association for Educational Communications and Technology, 1995).
Amir Levy and U. Merry, *Organizational Transformation: Approaches, Strategies, and Theories* (New York: Praeger, 1986).
Mathew B. Miles and Karen Seashore Louis, "Mustering the Will and Skill for Change," *Educational Leadership* (May, 1990), pp. 57–61.
G. Morgan, *Riding the Waves of Change: Developing Managerial Competencies for a Turbulent World* (San Francisco, CA: Jossey Bass, 1989).
Richard Tanner Pascale, *Managing on the Edge* (New York: Simon and Schuster, 1990).
Price Waterhouse Change Integration Team, *Better Change: Best Practices for Transforming Your Organization* (Burr Ridge, IL: Irwin, 1995).
David F. Salisbury and Daryl R. Conner, "How to Succeed as a Manager of an Educational Change Project," *Educational Technology,* 34, No. 4 (July–August, 1994), pp. 12–19.
Marshall Sashkin and John Egermeier, "School Change Models and Processes: A Review and Synthesis of Research and Practice," *U.S. Department of Education,* U.S. Government Printing Office (Washington, DC: GPO, October, 1993).
Dean Tjosvold and Mary M. Tjosvold, *Psychology for Leaders: Using Motivation, Conflict, and Power to Manage More Effectively* (New York: John Wiley & Sons, 1995).
Calhoun W. Wick and Lu Staton León, *The Learning Edge: How Smart Managers and Smart Companies Stay Ahead* (New York: McGraw-Hill, 1993).

Concerns Management
Gene E. Hall and Shirley M. Hord, *Change in Schools: Facilitating the Process* (Albany, NY: State University of New York Press, 1987).

Culture Change
T. E. Deal and A. Kennedy, *Corporate Cultures: The Rites and Rituals of Corporate Life* (Menlo Park, CA: Addison-Wesley, 1982).
Connie Goldman and Cindy O'Shea, "A Culture for Change," *Educational Leadership* (May, 1990), pp. 41–43.
Bruce Joyce, Carlene Murphy, Beverly Showers, and Joseph Murphy, "School Renewal as Cultural Change," *Educational Leadership* (November, 1989), pp. 70–77.
Jon Saphier and Matthew King, "Good Seeds Grow in Strong Cultures," *Educational Leadership* (March, 1985), pp. 67–74.
Grant W. Simpson, "Keeping It Alive: Elements of School Culture That Sustain Innovation," *Educational Leadership* (May, 1990), pp. 34–37.

NOTES

1. Changing our current education systems into more effective and efficient ones will require a major transformation. Case studies point out that it is sometimes easier to transform organizations by creating new systems or new units for which new personnel are recruited and trained than to renew an already existing system. Thus, it may be easier to create a chain of new, innovative schools that provide an alternative to the existing system than to renew the existing system.
2. These stages of change are described in Daryl R. Conner's *Managing at the Speed of Change: How Resilient Managers Succeed and Prosper Where Others Fail* (New York: Villard, 1993). Conner describes the stages of change from the point of view of the internal feelings and emotions of the people in the organization who are experiencing the changes. Other writers, such as Havelock (*The Change Agent's Guide*, 2nd ed., Educational Technology Publications, 1995) describe stages of change in terms of the actions or steps to be taken by the leader or change agent. Havelock's stages include *care, relate, examine, acquire, try, extend,* and *renew*.
3. Gene E. Hall and Shirley M. Hord, *Change in Schools: Facilitating the Process* (Albany, NY: State University of New York Press, 1987).
4. Another effective way to collect concerns is by using a telephone voice mail system where callers can leave anonymous messages in which they describe their concerns. The number should be advertised widely and people encouraged to call. The opening announcement on the voice mail system should say something like, "This is Superintendent Wolsley. I'm very glad you called this voice mail number to leave your comments or concerns about the _____ project. I'm sincerely interested in your thoughts about this and I value your input. Please be frank and explain your concern in as much detail as possible. I will listen personally to each message and will forward your comments to other appropriate individuals so they, too, can benefit from hearing your views. Again, thanks for taking the time to call."
5. Adapted from S. M. Hord, W. L. Rutherford, L. Huling-Austin, and G. E. Hall, *Taking Charge of Change* (Alexandria, VA: Association for Supervision and Curriculum Development, undated). For more information on analyzing open-ended concerns statements, see B. W. Newlove and Gene E. Hall, *A Manual for Assessing Open-Ended Statements of Concern About an Innovation*, Report No. 3029 (Austin, TX: University of Texas at Austin, Research and Development Center for Teacher Education), ERIC Document Reproduction Service No. ED 144 207.

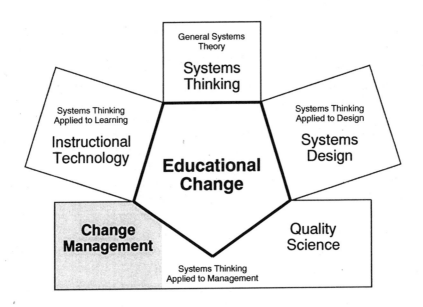

CHAPTER 9

CHANGE MANAGEMENT TOOLS

Change is not mysterious or inexplicable. We now understand that there are reliable and predictable stages and events that are inherent in the change process. Understanding these stages and events is essential to enjoying the benefits and avoiding the pitfalls that can be part of any organizational transition. Most books about organizational change discuss these known stages and events, but don't provide usable tools to help in the process. Understanding change in the abstract is useful and necessary, but one also has to have access to tools that can be used to replicate the actions and behaviors of those who have succeeded at implementing major change.

The tools that I present in this chapter have been rigorously field tested and validated over a period of time in various kinds of organizations. These tools are designed to bring structure and discipline to the process of change. The tools represent flexible and structured procedures to be followed in planning, implementing, and evaluating change. By using these tools, education innovators can better perform their roles as leaders and managers of change.

CHECKLIST FOR CHANGE

Thomas Harvey provides a twenty-step checklist for creating and controlling change in organizations. The checklist includes analysis, planning, implementation, and evaluation, and each step has intermediate steps. "Analysis" incorporates a description, a need, potential actors, the payoff (outcomes), understanding motivation, the resistance to change, the investment in change, and the culture. The category "Planning" covers strategies, the change environment, the scope of the change, and who will be affected by the change. "Implementation and Evaluation" includes time frames, observing the implementation, and analyzing the process. Checklist for Change provides a guide to implementing change in a simple, no-nonsense manner.

HARVEY'S CHECKLIST FOR CHANGE

Analysis
1. Description
2. Need
3. Potential actors
4. Payoff
5. Unfreezing
6. Resistance
7. Investment
8. Culture

Planning
9. Actual "changees"
10. Change strategy
11. Resistance strategy
12. Participation
13. Excitement
14. Change environment
15. Scope

Implementation and Evaluation
16. Advocates
17. Time frames
18. Monitoring
19. Action plans
20. Risk Analysis

—From Thomas R. Harvey, **Checklist for Change: A Pragmatic Approach to Creating and Controlling Change** (Needham Heights, MA: Allyn & Bacon, 1990). Used with permission.

MEASURING THE IMPLICATIONS OF CHANGE

All organizations can accommodate a certain level of change. However, if the change is too large or too fast, the people in an organization, and the organization itself, can go into overload. An organization is overloaded when the magnitude or speed of change is greater than the people's capacity to deal with it. Too many change initiatives happening concurrently can also send people into overload.

A number of scales exist that can be used to measure an organization's readiness for change. One such scale is shown on the next page in the insert "Measuring the Implications of Change Response Form." By having the people in an organization respond to this questionnaire, the leader or change manager can determine what the level of impact will be on the people in the organization. The results may indicate that the leader is dealing with a manageable level of change. Or the leader may conclude that there is reason for caution or concern about the magnitude of the change and may wish to reexamine the change plan.

MANAGER'S RESISTANCE ASSESSMENT

Rather than using a survey to collect perceptions and opinions about the change, leaders or managers can perform an exercise using their own personal responses or an estimation of likely responses by others in the organization. This three-step assessment tool (page 173) helps the manager or leader predict situations that will be created as a result of the change, estimate the degree of resistance that will be generated by the change, and reexamine plans for implementation.

COMMITMENT CHART

In any change effort there will be a number of key players whose commitment (see the table on page 174) will be needed for the change to occur. These players may be sponsors, advocates, agents, or targets. It is important for the leader to understand in advance the level of commitment of these key players and to make sure that there is a critical mass for the change. At the beginning stages of a change effort, there will likely be only a few early innovators. Most people in the organization may be fence sitters. The leader needs to know who to look to for support and early participation. This is where commitment charting comes in. Based on the work of Beckhard and Harris, commitment charting allows the leader to examine the motivation of the key players in the change on a scale from active antagonism to a positive intention to make the change happen.

MEASURING THE IMPLICATIONS OF CHANGE
RESPONSE FORM

Instructions: Considering your role as a participant in _____, we would like to have your opinion about how this change may affect you and the school. Please place a mark under the number that best describes your level of agreement with each statement below. Your opinions and reactions are critical to this survey.

Agree (1-5)	1	2	3	4	5	6	7	8	9	10	Agree (6-10)
1. There will be few modifications in the way I do my work.											1. There will be significant modifications in the way I do my work.
2. There will be little impact on the total school system during this change.											2. There will be substantial impact on the total school system during this change.
3. A sufficient amount of time has been allotted for this change to occur.											3. A insufficient amount of time has been allotted for this change to occur.
4. There is a clear understanding of the purpose of this change.											4. The purpose of this change is not clear.
5. I understand completely how participating in this change will affect me personally.											5. I have little understanding how participating in this change will affect me personally.
6. I have or know that I can acquire the skills necessary to participate in the change.											6. I do not have the ability to acquire the skills necessary to participate in the change.
7. I want to participate in the change.											7. I do not want to participate in the change.
8. Participation will cause few changes in my professional beliefs and values.											8. Participation will cause extensive changes in my professional beliefs and values.
9. Participation will not cause me to feel any differently about the people with whom I work.											9. I will feel differently about the people with whom I work because of my test participation.
10. Participation will not require me to learn new concepts and methods.											10. Participation will require me to learn new concepts and methods.
11. My daily routines will remain the same during my participation.											11. Participation will greatly alter my daily routines.
12. The use of electronic equipment will be a minimal part of this change.											12. The use of electronic equipment will be a significant aspect of this change.
13. Participation in the change will have no direct impact on my school's budget.											13. Participation in the change will adversely impact on my school's budget.
14. There will be no impact on local policy due to this change.											14. There will be significant impact on local policy due to this change.
15. There will be no impact on parents due to this change.											15. There will be significant impact on parents due to this change.
16. There will be no impact on students due to this change.											16. There will be significant impact on students due to this change.

—Adapted from ***Measuring the Implications of Change: The Florida Schoolyear 2000 Initiative***, *Center for Educational Technology, Florida State University (Tallahassee, FL: Florida State University, 1995) from text.*

Change Management Tools

> **MANAGER'S THREE-STEP RESISTANCE ASSESSMENT**
>
> **STEP 1:** If you believe that one of the following situations will be created as a result of your upcoming or current change effort, put a number next to it to indicate the degree of resistance that situation will engender (1 = low, 2 = medium, 3 = high).
>
> ___ Perceived threat to job security
> ___ Loss of expertise
> ___ Need to learn new skills
> ___ Shifts in influence, authority, control
> ___ Shifts in communication patterns
> ___ Loss of social status
> ___ Change in habits/customs
> ___ Limited understanding of the change and its implications
> ___ Low tolerance for change
> ___ Total
>
> **STEP 2:** Total the numbers in the left-hand column. If your total is 10 or below, you have a manageable level of resistance. If your total is between 10 and 20, you need to make a special effort to deal with a significant level of resistance. If you total is 20 and above, you should stop any current change effort and reexamine your transition management plans to be sure that you are managing these serious levels of resistance.
>
> **STEP 3:** Be sure to ask yourself:
>
> What are the implications of these areas of expected resistance?
> What can be done to mitigate the effects of resistance in these areas?
>
> —*From The Price Waterhouse Change Integration Team,* **Better Change: Best Practices for Transforming Your Organization** *(Burr Ridge, IL: Irwin, 1995). Used with permission.*

On the change commitment chart, the 'Xs' indicate where that person's commitment is at present, the 'Os' indicate where their commitment needs to be to ensure that the change is successful. The length of the arrows indicates the amount of work that must be done to shift individual commitment to the needed level.

Player 1, for example, is currently placed in the 'Let it happen' box but his commitment must move to the 'make it happen' level for the change to be successful. Player 10 is currently antagonistic to the change and, while it is not essential that he drive the change, he must somehow be persuaded to let it happen. Player 3, who is currently pressuring for the change, must be

COMMITMENT CHART

Key players	Antagonistic	No commitment	Let it happen	Help it happen	Make it happen
1.			X		→O
2.			X	→O	
3.			O ←		X
4.			O ←	X	
5.				(XO)	
6.			(XO)		
7.			(XO)		
8.			(XO)		
9.		X		→O	
10.	X		→O		

—Adapted from Richard Beckhard and Reuben T. Harris, **Organizational Transitions** *(Reading, MA: Addison-Wesley, 1987)*. Used with permission.

pulled back and persuaded to let go and let it happen. By making this *his* change rather than the organization's change, he is jeopardizing the effort. Where the 'Xs' and 'Os' coincide in the same box, as in the case of players 5, 6, 7, and 8, the leader only has to reinforce the appropriate commitment. More attention must be directed to those who are not yet demonstrating the most effective level of commitment.

CLARIFYING THE CHANGE

Whenever you undertake major change, you can expect some confusion and ambiguity. Confusion and ambiguity are part of change. Clarifying the change as much as possible will help to decrease the amount of confusion and misunderstanding that will be present during the change process. The change clarification tool shown in the following insert can help the leader or team clarify things about the change that might be sources of confusion or misunderstanding later on. The leader or team should seek to answer all of the questions in the clarification tool before feeling ready to announce or initiate the change. If the leader or team cannot answer these questions, the change is not well understood. Further exploration should be conducted before initiating the change.

CHANGE CLARIFICATION TOOL

1. Name the change.
2. Briefly describe the change.
3. List primary components that must be in place, or any prerequisites required for the success of the change.
4. List the technical objectives of the change.
5. List the human objectives of the change.
6. Outline the time frame for the change including preparation, execution, and evaluation.
7. Outline budget constraints of the change including equipment, service, training, user expenses. Identify other constraints.
8. Identify what consensus decisions are necessary to support and drive this change. Who will make these decisions?
9. Identify sponsors who must legitimize this change.
10. Identify why the success of this change is important to the sponsor.
11. Identify why this change is important to the customers or clients.
12. Identify who the primary individuals are who will be affected by this change. Identify who will have personal implications and what those implications might be.
13. Identify why this change is important to the target. Is there a benefit to the target?
14. Identify messages for the targets and means to deliver these messages.
15. Determine potential issues and answers that concern the targets.
16. Determine anticipated resistance and how to diminish it.
17. List responsibilities of all change agents.

—Adapted from *Change Clarification for Operational Tests, The Florida Schoolyear 2000 Initiative* (Tallahassee, FL: Florida State University, Center for Educational Technology, 1995) from text. The Change Clarification for Operational Tests form was developed with help from ODR, Inc.

CHANGE TIME SCALE

Even though change is frustrating and in some ways erratic, there are still known stages and phases of change than can be described and predicted. Simply understanding and communicating information about the process of the change can be helpful to all in the organization as they acknowledge frustrations and wait for the change to produce results. One tool that is useful in communicating the process of change is the "change time scale."

The change time scale demonstrates the inevitable fall-off in productivity as change occurs. Initially, productivity may increase due to the attention given the new innovation, with its announcement by slogans, bands, and fireworks. But, inevitably, there is a fall-off in productivity as

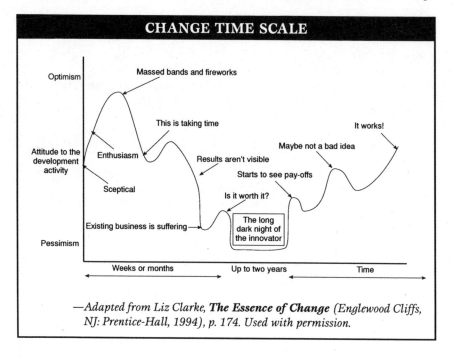

—Adapted from Liz Clarke, ***The Essence of Change*** *(Englewood Cliffs, NJ: Prentice-Hall, 1994), p. 174. Used with permission.*

people get into the nuts and bolts details of implementing the change. It may appear that absolutely nothing is happening, that the change is taking too long, and that there are no visible results for what is perceived as a long time. During this period, enthusiasm dwindles, cynicism spreads that this is yet another 'flavor of the month' initiative, top administration may back away from the change because it appears to be failing. They may even abandon the change and launch something else in its place, again with fanfare, bands, and fireworks (only to see that fail as well). If top administration understands the change time scale, they will realize that everything looks like a failure half-way through, and they will stay with the change during this period.

If they persevere and maintain enthusiasm and commitment, they can lead the change through the "long dark night phase" to the point where they start to see pay-off. Understanding this time scale of change is critical to leaders of change. Communicating the time-scale to others in the organization can prepare everyone for the long dark night and increase the probability of success.

LEVELS OF USE SCALE

Before drawing conclusions about the positive or negative effects of a change, we want to be sure that the change has actually been implemented.

LEVELS OF USE SCALE

Each level on the Levels of Use Scale involves behaviors that are or are not taking place in relation to the change. Because Level 1 represents the lowest level of use, it is easier to understand the levels if you start reading the scale at the bottom (Level 1) and move upward to Level 8. As you move up to higher levels, each level involves more positive behaviors related to the change.

LEVEL 8	RENEWAL	State in which the person reevaluates the quality of the change, seeks major modifications of or alternatives to present operations to achieve increased impact on students, examines new developments in the field, and explores new goals for self and the organization.
LEVEL 7	INTEGRATION	State in which the person is combining own efforts with those of colleagues to achieve a collective impact on students within their common sphere of influence.
LEVEL 6	REFINEMENT	State in which the person varies the change to increase the impact on students within immediate sphere of influence. Variations are based on knowledge of both short- and long-term consequences for students.
LEVEL 5	ROUTINE	Operations are stabilizing under the change. Few if any adaptations are being made to the change. Little preparation or thought is being given to improving the change or its consequences.
LEVEL 4	MECHANICAL USE	State in which the person focuses most effort on the short-term, day-to-day aspects of the change with little time for reflection. The person is primarily engaged in an attempt to master the tasks required by the change, often resulting in disjointed and superficial use.
LEVEL 3	PREPARATION	State in which the person is preparing for the change
LEVEL 2	ORIENTATION	State in which the person has recently acquired or is acquiring information about the change and/or has recently explored or is exploring its value and its demands upon him or her personally.
LEVEL 1	NONUSE	State in which the person has little or no knowledge of the change, no involvement with the change, and is doing nothing toward becoming involved.

—Adapted from Gene E. Hall and Shirley M. Hord, **Change in Schools: Facilitating the Process** *(Albany, NY: State University of New York Press, 1987).*

The Levels of Use Scale on the previous page is an instrument for assessing the level of implementation of a change. By determining the level of use of an innovation, we can assure that the change has been adopted, effectively implemented, and institutionalized. Also, use of the scale prevents us from drawing conclusions about the change without assessing whether it has been fully or only partially implemented.

Eight levels of use have been identified and operationally defined. Each level is described briefly in the insert. There is a typical progression in a person's movements through the levels of use. The progression is not lock-step, however, because people often jump several levels at a time. This progression assumes, of course, that the change is appropriate, the organization leaders and other change agents do their job in managing change, and appropriate time, training, and support are provided to the persons involved in the change.

Stages of Concern, described in the previous chapter, can be useful in tracing the feelings and perceptions of individuals as they experience the change and may reveal why an individual or group has not moved up to a higher level on the Levels of Use scale.

OTHER SOURCES FOR CHANGE MANAGEMENT TOOLS

ODR Inc.

ODR, Inc. is an international consulting firm that specializes in helping organizations manage change. ODR offers a number of proprietary diagnostic, planning, and implementation tools that are designed to enable their clients to be successful in implementing large-scale organizational change. No other firm that I know of has monitored change in more organizations. By observing, recording, and analyzing the behavior of managers as they attempt to implement major change, they have been able to identify the perspectives and behaviors that contribute to their success or failure. These patterns eventually constituted a structure for describing the process of successful change and for developing the tools listed in their catalog. Most of ODR's clientele are Fortune 500 companies and their resources are priced accordingly. Additional information about these tools and resources can be obtained from ODR's *Practitioner's Guide to Diagnostic Services and Sequence for the Comprehensive Application of the Managing Organizational Change Methodology*. See the insert on the next page.

Price Waterhouse

The Price Waterhouse Change Integration Team is a group of nationwide practitioners whose focus is on large-scale change projects in corpora-

SELECTED TOOLS AVAILABLE FROM ODR, INC.

Situational Analysis Tools
Implementation Problems Assessment
Predicting the Impact of Change
Sponsor Evaluation
Synergy Survey
Change Resistance Scale
Change Agent Evaluation
Culture Assessment

Planning and Communication Tools
Preliminary Implementation Plan
Organizational Change Implementation Plan
Project Analysis
Constituency Analysis
Statement Development
Announcement Plan

Application Tools
Role Mapping Application Tool
Pain Management Strategies: Sponsor
Conflict Management Guidelines

Tools for Technological Change
Technological Change Resistance Scale
Predicting the Impact of Technological Change
The Culture Assessment for New Technology

—*The above list of tools was selected from a recent ODR order form. The order form and their most current list of tools can be obtained from ODR, Inc. Building 16, 2900 Chamblee-Tucker Road, Atlanta, GA 30341-4129. Toll free: 1-800-CHANGE-U; Fax: 404-455-8974.*

tions. They have a number of proprietary change management and analysis tools for use with their clients, including a facilitator's guide and materials for a change readiness workshop. Many of these methods and tools would be applicable to schools.

Andersen Consulting

Andersen Consulting's Change Management Group works with organizations (including schools) to effectively manage change. Andersen Consulting has gathered and codified the best worldwide practices. It offers a field-tested methodology for helping organizations change.[1]

FOR FURTHER READING

Michael H. Annison, *Managing the Whirlwind* (Englewood, CO: Medical Group Management Association, 1993).

Richard Beckhard and Reuben T. Harris, *Organizational Transitions* (Reading, MA: Addison-Wesley, 1987).

Richard Beckhard and Wendy Pritchard, *Changing the Essence: The Art of Creating and Leading Fundamental Change in Organizations* (San Francisco, CA: Jossey-Bass, 1992).

Lance A. Berger and Martin J. Sikora, *The Change Management Handbook: A Roadmap to Corporate Transformation* (Burr Ridge, IL: Irwin, 1994).

Liz Clarke, *The Essence of Change* (Englewood Cliffs, NJ: Prentice-Hall, 1994).

Gene E. Hall and Shirley M. Hord, *Change in Schools: Facilitating the Process* (Albany, NY: State University of New York Press, 1987).

Thomas R. Harvey, *Checklist for Change: A Pragmatic Approach to Creating and Controlling Change* (Needham Heights, MA: Allyn & Bacon, 1990).

Ronald G. Havelock with Steve Zlotolow. *The Change Agent's Guide,* 2nd ed. (Englewood Cliffs, NJ: Educational Technology Publications, 1995).

Price Waterhouse Change Integration Team, *Better Change: Best Practices for Transforming Your Organization* (Burr Ridge, IL: Irwin, 1995).

John W. Thomas, *Making Changes: A Futures-Oriented Course in Inventive Problem Solving,* Teacher Guide (Philadelphia, PA: Research for Better Schools, 1981).

NOTE

1. For more information about Andersen Consulting's Change Management practice, contact Andersen's Change Management for the Americas at 216-781-3580.

PART V

The Fifth Technology: Instructional Technology

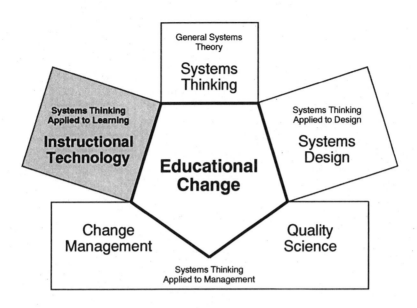

CHAPTER 10

INSTRUCTIONAL TECHNOLOGY

The world has changed from the days when instruction consisted always of lecture, seat-work, and memorization, and when the only tools that the students or teacher used were the textbook, chalk, pencils, and paper.

Perhaps the most promising vehicles for changing schools are the same computer and video technologies that are changing the industrial economy into the knowledge economy.

The four technologies discussed in the book so far—systems thinking, systems design, quality science, and change management—are part of a set of five fundamental technologies for improving education. The fifth technology of the set is instructional technology.

Instructional technology is part of this collection of five essential technologies because without it, no substantial reform, redesign, or restructuring of education is possible. The world has gone far beyond the days when instruction and learning was connected only with presentation by the teachers followed by seat-work, and when the students and teachers used only textbooks, chalk, pencils, and paper as tools. Businesses and large corporations that conduct education and training programs for their workers and executives have caught onto instructional technology as an effective way to make learning efficient, enjoyable, and more cost-effective than traditional methods. U.S. businesses spent more than $50 billion on education and training in 1994.[1] Many of these companies provide management and skills training and also remedial and basic education. They depend on the techniques and practices of instructional technology to provide the most effective learning. Schools, unfortunately, lag way behind in their application of this powerful and effective technology for improving learning and instruction.

WHAT HAVE WE LEARNED?

Prior to about 1960, our knowledge about instructional techniques had not advanced to the point where more effective alternatives to traditional instruction (a teacher in front of a class of 20–30 students) could be proposed. Today, however, the situation is different. Our knowledge has now evolved to the point where more effective methods of instruction are known.

The main thing we have learned is that teaching and learning can be treated scientifically. We can now determine with relative exactness what a student will need to learn to be able to perform well in a particular area. We can diagnose learning problems or gaps in a student's understanding of the subject matter to be able to help that student with his or her specific learning problem or need. We also have better ways of measuring students' understanding so that we don't waste time teaching them something they already know or risk teaching them at a level they are not prepared to understand. We have better ways of helping students to be engaged and motivated to learn by tailoring instruction to their individual learning levels and styles. We know how to present instruction so that it will gain their attention, stimulate recall and learning, and allow them to practice and receive feedback. We know how to observe the learning process to pinpoint where the instruction is weak so that we can correct it for the next student or group of students.

Instructional technology gives us scientific techniques for producing improvements in the way students learn and in the way we present instruction to those students. It seeks to make learning effective for most students. It does that by including actions and procedures that allow us to:

1. Describe what the instruction should accomplish (what should the student be able to do after the instruction is complete?)
2. Determine what the student can already do in relation to the final objective.

3. Develop, try-out, and revise instruction that will allow the student to perform well.
4. Implement the instruction either with individual students or in group settings.
5. Use the best way of delivering the instruction to the students (teacher presentation, computer, videotape, textbook, peer-tutoring, etc.)
6. Evaluate how well the instruction worked with those students.
7. Improve the instruction so that it will work better the next time.

Instructional technology, then, is a scientific process used to develop, fieldtest, and improve instructional products and programs. Unfortunately, most programs in place in today's schools were not developed in this way. A number of untested fads sweep through the profession each year, gathering authority by the number of schools using them, not by proven gains in student learning. This unscientific approach to education squanders the time, money, and effort of those trying to improve learning for students.

Most teachers are not aware of instructional technology. They are only aware of the many untested innovations that become popular each year. Often, these innovations are no better than what was done in the past. The enthusiasm and energy of new teachers rapidly changes to dissatisfaction and cynicism when they are caught up in a cycle of untested innovations that merely create change, but do not yield improvements in student achievement.

A recent survey conducted by the National Center to Improve the Tools of Educators found that 76% of the teachers surveyed believed school reforms had not benefited the students who need help the most. The survey results indicate that teachers want facts, not fads, to govern the selection of new programs and materials. Teachers responding to the survey said that new innovations should be scientifically tested before using them in the classroom and schools should only use the ones that can be shown to produce results.

According to an Education Product Information Exchange Institute study, approximately 90% of the 309 billion hours Americans spend in school involves the use of textbooks, activity guides, films, or other materials or preproduced programs.[2] These tools and practices influence between 75% and 90% of what goes on in classrooms. If these materials are not of high quality, students will not learn as quickly or as well. If you have tried to teach with bad materials, materials that are wrong, ambiguous, or too difficult, or have tried to help your own child with his or her homework when the textbook or other materials do not provide enough practice or review, you know the negative influence that these materials have on learning.

Instructional technology, as developed over the last 30 years, is a process used primarily to develop instructional programs and products such as print materials, computer-assisted instruction, and televised instruction. Research and experience has shown that programs and materials developed using the instructional technology process are very effective. This is because these materials have been tested with actual students, and have been revised several times until they are shown by this process to be highly effective. Only after a careful process of research, experimentation, and verification does a new approach gain the status of an accepted instructional practice or product; those that fail are discarded. Instructional technology can shield teachers and students from unnecessary failure and wasteful expense due to bad instruction. It costs no more to use a tested and validated product or program than it does an untested innovation. But unproductive innovations not only waste money—they also misuse the energy of school personnel and the hard work of children trying to excel.

My intent in this book is not to teach you how to become an instructional technologist. This would not be possible in one single chapter of one book. I do present some of the major instructional technology tools and techniques in the next chapter. That chapter will show you some of the basic tools and methods instructional technologists use in their work. But keep in mind that most professional instructional technologists receive advanced training in the many universities that now offer programs in this relatively new field. Also, most instructional technologists are not teachers. They are professionals who spend the majority of their time planning and developing instruction to be used by others. They generally spend a relatively small portion of their time actually interacting with students. My purpose, then, in this chapter and the next, is simply to acquaint you with the process of instructional technology and to convince you that it is a necessary component for effective educational reform.

INSTRUCTIONAL TECHNOLOGY AND ELECTRONIC TOOLS

Instructional technology consists of the best practices currently available for modifying human capabilities. It includes practical procedures for using computers and multimedia devices either to deliver the instruction directly to the student or to supplement portions of what is presented verbally by the teacher. Our current system depends far too much on teacher presentation, and group (rather than individual) learning, and does not make serious use of new electronic learning tools. The computer and information age make a more modern educational enterprise necessary. The electronic tools of today and tomorrow have the capability of providing educational activities in ways we did not think possible a few years ago. And the power of these tools increases daily, while their cost continues to decline dramatically.

Instructional Technology

We now have at our disposal practical, cost-effective means for educating students through multimedia learning systems that are superior to many old methods. Working on their own, or under the supervision of a teacher, students can download a file that contains the lesson appropriate for them that day, based on their own progress. They can control the pace, sequence, level of difficulty, and the amount of practice that they want or need in a given area.

If these tools are so powerful, why then haven't they already had a major impact on improving student productivity and learning in schools? The reason is that high-tech learning tools have been applied only at the margins of a school's activities. Where schools have made substantial investment in purchasing new electronic learning tools, they have applied these new tools only as separate components—something added onto an already obsolete instructional process. Many, including myself, have tried to apply instructional technology along with computer learning tools in schools without changing the other aspects that impact on the instruction. It doesn't work and it won't work! Experience and a mountain of research have revealed the difficulty of using high-tech instructional resources effectively in today's schools.[3] These new electronic learning tools and resources provide the means of allowing instruction to be truly "individualized"—students can go at their own learning speed, select learning tasks, request additional practice, test themselves, and explore areas of individual interests. Unfortunately, such an individualized approach is incompatible with schools as they are currently operated. This is because schools operate around the concept of forming learners into groups and having a single adult deliver instruction to the group as a whole. Those who already understand the material are bored, while those who are slower flunk out. Proper use of electronic learning resources is simply incompatible with this approach. Trying to implement innovative approaches in this kind of a context is like swimming up a waterfall. The restraining forces exercised by the institutional barriers and incompatibilities are so overpowering that there is no possibility of success.

One of the major factors that work against the effective use of electronic learning tools in our schools has been described by Larry Cuban of Stanford University as the "tradition of teaching." The tradition of teaching relates to the fact that very few of those entering the teaching profession have ever experienced anything other than the delivery of instruction by the teacher. This is usually true not only of the years the new teacher spent as a student but also of the time spent in a teacher-training program. These programs train prospective teachers to function in the traditional classroom lecture/seatwork model of schooling—because that is the only model that really exists out there. Electronic learning tools cannot be used effectively if they are used only to automate the lecture/seatwork model.

THE CLASSROOM IS OBSOLETE

We sometimes hear the phrase "using technology in the classroom." In fact, this idea is counterproductive simply because the classroom *is* a technology—a technique designed for a learning age when the teacher was the dispenser and source of all knowledge. To facilitate this role, we designed the self-contained classroom, put the teacher's desk and a chalkboard in front, sat students in rows, and organized them into groups that met for the same amount of time each day and moved on to the next class at the sound of a bell. This technique (or technology) served us very well for a period of time. But today's computer and telecommunications revolution is spawning an educational crisis. The classroom method of educating students has become obsolete. It is simply not able to provide the individualized, high-speed, customized education that is now possible and essential.

Also, the amount of new knowledge is exploding so rapidly that no individual teacher can keep up with the flow of new developments in a given field of study. It has been estimated that the volume of new knowledge in the world is doubling every three years. What this means is that 95% of all the knowledge that will be available in the year 2004 will be discovered between now and then. Most fields of knowledge are already expanding so quickly that no single teacher can hope to keep up with new developments. As new knowledge explodes, old knowledge becomes more valuable as well. If students are to make sense of the vast amount of new information that will be available to them, they will need to be culturally and classically literate. If students are not firmly grounded in the cultural traditions, literature, and great thoughts and ideas of the past, they will not be able to understand the current issues, express their viewpoints, and act accordingly.

We once lived in an age when only teachers had access to the knowledge-base of the past and when oral presentation to groups of students was the most efficient way to pass that knowledge-base on to the rising generation. New information was only available to teachers, and it made sense to go through a chain in which this information was passed first to teachers, then from teachers to students. The information and telecommunications revolution is changing all that. Today, students can access the knowledge-base directly. In the information and communication age, the role of the "teacher" will change from one of disseminating knowledge to one of motivating, advising, and managing the students' learning. Teachers will introduce children to well-designed learning resources as the children become ready for them. Computerized multimedia resources (including interactive computer, videodisc, Internet, and CD-ROM systems), peer tutors, projects, and learning labs will be used to provide access to most skills and knowledge. Teachers can then do more of what only people can do—develop strong personal relationships with students, diagnose learning problems and prescribe activities to remedy them, and focus on the really difficult, really important aspects of learning—things like how to think and write well, how to solve complex problems, how to work as an effective member of a team, how to relate to other people, and so on. These things cannot be done by computers or books or videos, and they can move education to a higher level.

That the classroom (as the basic setting for instruction) has finally become obsolete should not be a surprise. Everything eventually reaches a stage of maturity and obsolescence. Witness the steam engine, the kerosene lamp, and the horse-drawn carriage. Things become obsolete when they can be replaced by something much better, faster, and cheaper—something that better meets people's needs. The classroom and group learning may still have a place for certain kinds of learning or subject matter. Discussion groups and seminars are especially useful in areas such as literature, and, of course, for music and other group performance activities. Certainly students like to be together at times, and teachers can often do valuable things in a group setting; but with a highly technological, rapidly changing, information-oriented society, the present structure of our educational system will become more and more inadequate.

Other sectors in our society—business, health care, manufacturing, finance—have had to turn to modern electronic tools, communication networks, and other information sources to maintain viability in the modern technological world. Schools will have to do the same thing. Unless the education enterprise can rethink and redesign the way it does business to fit a more modern technological world, it will find itself at the obsolete edge of human affairs.

The current model of education, which relies so heavily upon the tradition of teaching and group instruction, simply cannot increase its effectiveness much beyond current levels. Educator Ted Sizer is one of those who has described the existing classroom model of teaching as outdated: "I don't think we've gotten to the heart of the problem yet. The problem is the existing system. And until we face up to that fact—that the existing system has to change—we're not going to get the kinds of changes that everybody wants."[4]

THE TECHNOLOGY INVESTMENT DISCREPANCY

Schools have not looked to computer systems and new multimedia devices as a major means for improving productivity, efficiency, and effectiveness. Some schools have, certainly, purchased computer workstations and other modern equipment. However, a study by the U.S. Congress' Office of Technology Assessment (OTA)[5] showed that the computers that had been purchased were not intensively used: Some were used only for remedial instruction by a small group of students, others for teaching programming in computer science classes. Furthermore, not all the students used computers, and those who did spent little more than one hour per week with them. Only a little more than half of the high schools in America provide enough computers to maintain a ratio of more than 1 for every 30 students. Junior high schools and elementary schools provide even fewer computers per student. Even when there are computers, they are generally stand-alone workstations rather than linked together into a computer network.

If computers and computer networks were being used properly, they would be integral to the day-to-day, hour-by-hour work of every teacher and student. This is certainly not the case today. If computers were to be eliminated entirely from our schools, instruction would go on without anyone noticing that anything was different. Presentation of instruction is still done largely by the teacher, and students are still paced through instruction in groups. Instructional media, such as television or videodisc, if used at all, are used only as presentation aids for the teacher.

If we were using these new tools as they should be used, it would be difficult to think of schools functioning without them. Try to imagine a modern bank operating without computers. When the computers in a bank go down, the employees switch from standard operating procedures (which rely on the computer system) to tedious and labor intensive (and therefore expensive) temporary back-up procedures that rely on paper forms. It should be the same in schools. What we have in schools today is analogous to a bank still trying to operate in the pre-computer age.

The reason for this is that education has by far the lowest investment in electronic equipment of any major service industry. According to the same OTA study,[6] the average level of investment in this type of equipment for U.S. companies as a whole is about $50,000 of capital investment per job. That means that most companies spend about $50,000 to purchase the tools, equipment, and facilities that each worker or executive uses. Some high-tech industries invest as much as $300,000 or more in equipment, tools, and facilities for each worker. Even labor-intensive "service" businesses provide at least $7,000 to $20,000 worth of equipment and facilities for each employee. Education, however, spends a measly $1,000 for tools and equipment per employee. So when it comes to investment in tools and resources, the education industry is far behind other enterprises in empowering its employees with computers and other electronic tools.

The situation is actually worse if one considers that most of the work of education is, after all, done by the students—not the teachers. If we consider the students as employees, in addition to paid staff, the education sector's small capital investment of $1,000 per employee becomes a pathetic $100 per worker if "worker" also means "student."[7] So here we have the nation's largest information industry—education—missing out on the explosive revolution of knowledge and information that has transformed almost every other sector of our society.

Educators, those in charge of school budgets, will say that there just isn't enough money to buy high tech tools, electronic learning resources, or computer networks. This is because approximately 85 percent of the total education budget goes toward salaries. In other words, almost all of the money spent in education goes to pay the salaries of teachers, administrators, and other school, district, and state employees. Very little of it is left over to pay for capital investment (buildings, desks, blackboards, and computers). The cost of supporting such a labor-intensive method of education has been going up dramatically for the past several decades [see insert "Public School Spending (1945-92)"] and there is no indication that the trend will slow down if we continue to operate in the same way. A truer statement about the prohibitive "costs" of computer and multimedia learning devices would be: "There is not enough money *not* to buy them."

The purpose for using electronic learning tools, computer networks, and multimedia devices in schools is to make the educational process more

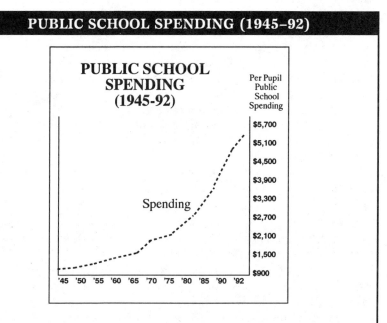

The cost of public education has increased at about the rate of 40% each decade since World War II. Costs have gone from about $1,000 per student in 1945 to over $5,000 per student in 1990 measured in constant dollars.

—*The Digest of Education Statistics*, National Center for Education Statistics, U.S. Department of Education (Washington, DC).

effective and efficient—that means better, faster, and cheaper. Most businesses have discovered how these tools reduce costs and boost productivity; most schools, though, have not. Instead of thinking of these devices as an "added-on" expense, we must think of investment in them as an alternative to spending that money in other ways.

Boosting productivity means that more output is produced for less money. The only way to accomplish this in education is to reduce the amount of money spent on salaries, and instead, spend more of it on tools and equipment that allow people to function better and more efficiently. Differentiated staffing, where adult personnel with different levels of qualifications and pay grades are employed in the school, is one means to reduce personnel costs. This may mean that the more experienced, highly paid teachers serve as leaders of other teachers and paraprofessional staff. Such differentiated staffing models could result in significant fiscal savings. Some schools that have tried differentiated staffing arrangements have found that they can reduce the adult to student ratio and still save costs. In other words, there are more adults helping the students, but not all are paid at the professional teacher level. Such restructuring of the roles of personnel will be necessary to allow for a different distribution of resources.

What schools have ignored is the fact that the costs of paying for computer equipment and other electronic learning tools is much less, in the long run, than are the ever-increasing costs of high-credential personnel. Education costs have continued to increase over the past several decades and they will continue to increase as long as we continue paying for more and more teachers to deliver instruction in outdated ways to groups of students sitting in a traditional classroom. At the same time, as education costs are going through the roof, the cost of producing computer-based learning materials, lessons, exercises, and multimedia presentations has continued to decrease. Today, the cost of an hour of computer-based instruction is well below the cost of paying a teacher to make the same presentation. So using computer-based learning is much *less* expensive than traditional instructional methods.

Business and industry, which invest substantial time and effort in the training of their employees, have discovered the sizable savings and increase in effectiveness that can result from the use of computer-based instruction. Approximately 30 percent of the more than $50 billion that employers invest annually in employee training is spent on computer-based instructional systems.[8] That is over seven times more in one year than U.S. public schools spent on computer-based instruction in the last decade! Schools should be leading the way in applying better instructional techniques, yet they are not even in the ballpark when it comes to the share of their budgets that they invest in computer-based instruction.

Private sector companies that use computer-based and other high-tech tools have found that these tools are very effective and affordable. Over two decades of research shows that computer-based instruction produces about

30 percent more learning in 40 percent less time and at 30 percent less cost than conventional classroom instruction.[9]

It should be noted that these long-term averages in savings and increased learning resulting from computer-based instruction were obtained from the use of computers and software that isn't even state-of-the-art anymore. By the time you read this, computer power will have doubled or tripled and the quality of software will have improved dramatically. We now are seeing the rapid emergence of full-feature multimedia products that add full-motion video and sound to the desktop computer. With these kind of systems, students can experience instruction that is truly 'interactive,' permitting them to view realistic pictures, practice skills in simulated settings, observe demonstrations, and receive feedback on their own performance.

INCENTIVES AND REWARDS

Schools have invested little in productivity enhancing tools primarily because there is little incentive to do so. Because the school system is primarily government controlled and bureaucratic, there is an absence of the basic competitive market forces that pressure other organizations (like businesses and hospitals) to invest in innovations to improve productivity and better serve their clients and customers. Private schools haven't had to invest much in computer tools either—since they don't get much competition in this area from the public schools.

The key factor in improvement is not altruism, good-will, or quality programs. It is competition! Competition is what makes corporate and business executives accountable for costs and results. The lack of competition in education makes for a lack of accountability in these same areas. For this reason, programs to install business-like competition in the government-run monopoly of public schools should be given every opportunity to work. There is no reason to believe that competition cannot do for schools what Federal Express and UPS have done for the U.S. Postal Service. It would be hard to find someone who honestly believes that American-made cars are not of higher quality today due to competition from Japan, Korea, and Germany, or that telephone service would be as customer oriented (in terms of pricing, features, and services) without the breakup of AT&T. Without competition, we have mediocrity. Without competition in the school system, we will continue to have mediocre schools.

Because public schools are a government-run monopoly, there is really no external economic incentive for educators or administrators to adopt innovations or improvements. Schools do not offer any substantive reward, or even opportunity, for professional staff to adopt productivity tools for themselves or for their students. In fact, there are benefits to operating inefficiently (so as to increase the organization's budget). So the school benefits by teaching a lot of remedial courses, starting drop-out prevention pro-

> ### EDUCATION: AMERICA'S COLLECTIVE FARM
> Continual attempts to inject technological innovation into American schools through subsidized experimental, pilot, and demonstration projects or top-down bureaucratic mandates have failed as thoroughly as similar initiatives in the Soviet state agricultural system...In essence, the public school is America's collective farm. Innovation and productivity are lacking in American education for basically the same reasons they were scarce in Soviet agriculture: absence of competitive market forces.
> —Lewis J. Perelman, **School's Out: Hyperlearning, The New Technology, and the End of Education** (New York: William Morrow, 1992), p. 225.

grams, labeling more and more kids as "learning disabled," and inflating costs rather than saving taxpayers' money by increasing efficiency, productivity, and results. These disincentives have been highly effective in preventing technological change and innovation in education.

Tax-supported schools often aren't compelled to embrace reforms because there are no serious consequences when students perform poorly. For good or bad, it is human nature that people will work harder, be more creative, innovative, and productive, if they stand to benefit personally from their efforts. Only when people stand to gain do they become very accountable for what they do or don't do. If a person's personal income and quality of life can be improved by increasing his or her productivity and effectiveness, the individual will put much more effort, creativity, and thought into the process.

What this means is that we need to create a system whereby teachers can expect to benefit personally and financially by being more productive both individually and collectively. This could be done in a number of ways. For example, several teachers/administrators band together to form a "cluster." The income of the teachers in the cluster is based on the number of students they attract and the cluster's gross income. This creates an incentive both for individual productivity and performance and for group excellence and cooperation. It also makes the cluster very accountable for what it does or doesn't do. If a cluster does a bad job of meeting parents' expectations and requirements, its income will fall, as will the income for all of the teachers in the cluster.[10] With these types of incentives, schools would have the motivation to invest in high-tech learning resources.

THE PRODUCTIVITY PARADOX

Spending a larger part of the education budget on electronic machines will not, in and of itself, improve productivity. To obtain real productivity increases through the exploitation of new tools and modern equipment, schools will need to take a lesson from what has occurred in business and industry. Even with high levels of investment in computers, robotics, and automation, businesses and industries did not, at first, experience large increases in productivity. During the last decade, manufacturing and service industries have invested more than a $1 trillion in computers and related devices, yet productivity gains in the manufacturing and service sectors has run at about two percent and 2.4% respectively during the same time period.[11]

The question of why such a large investment in modern equipment has not produced larger gains in productivity has become known as "the productivity paradox." The explanation seems to be that investments in high-tech tools do not translate into improvements in productivity if the tools are used simply as a substitution for the old processes. When desk-top computers were first introduced, they functioned primarily as glorified typewriters and adding machines. They allowed people to complete the work faster and more efficiently, but did not change the nature of the work being done. Only when people started to rethink the nature of the job and completely overhaul the process did we begin to see spectacular productivity gains.

A prototypical example is that of a large medical-products manufacturer that installed a new high-capacity computer system into its assembly-line operation. The system was capable of much more productivity and customization than their current equipment. Yet neither the managers nor the operators were comfortable with the new system. As a consequence, they used the system basically for the same tasks and in the same way as they had used the older equipment. As a result, productivity barely increased, making the firm's $2 million investment in the new computer system seem unjustified. Ultimately, they tried a different approach. They set up the new machine in another portion of the plant. New managers and employees were chosen for the operation. They redesigned the entire production process, changed incentive structures to reward workers for creative ideas, and allowed workers and management to exploit the new system to its full capacity. The results were spectacular, and the company is now trying to incorporate these innovations into the main portion of the plant.[12]

Overhauling work processes, changing the nature of the task, and rethinking processes is known as "reengineering." It has become the watchword in large manufacturing operations and has recently begun to penetrate the service sector. We're now seeing lots of restructuring and rethinking of jobs and tasks in telecommunications, retailing, airlines, financial services, and government agencies. These organizations, to attain needed productivity increases, find themselves "flattening" their organizational structures, decentralizing decision-making, outsourcing more of their work, and allowing people to work at home. We also see the emergence of "virtual corporations"—small, sometimes temporary companies formed by one or two individuals (often living in different geographic locations) that, through electronic communication, coordinate production or delivery of goods or services and constantly change and adapt to changing consumer demands.

If our education systems are to prosper, we're going to have to reorganize our ways of working and learning so that we can fully profit from new electronic tools and equipment. This cannot be done by just using these tools as a supplement to the traditional teacher/classroom model. New and different kinds of jobs and staffing patterns will need to emerge for teachers and other school personnel. Students can begin to undertake some aspects of the instructional activities, such as tutoring their peers, helping to maintain the school's computers, leading teams, and assisting teachers. Schools will need to accommodate electronic groups, on-line learning at home, part-time students, full-day operation, and other such practices that will allow modern learning tools to be fully exploited.

LEARNING IN THE ELECTRONIC WORLD

In the future, schools may consist of learning labs designed more along the lines of a modern automated office environment rather than classrooms designed for teacher delivery of instruction. There may be labs designed for traditional, discipline-oriented learning (such as biology) or for cross-disciplinary, problem-oriented activities such as space exploration, natural resources, and creating inventions. The teachers would be responsible for helping students put together an individualized program of study that represents the core curriculum requirements and also considers each student's interests and goals. The teacher and the student monitor the student's progress and use acceptable norms, benchmarks, and standardized tests as reference points. Resources would be allocated to the labs based on usage.

Some labs may be mobile units that travel around to schools, shopping malls, museums, public libraries, parks, or businesses. (For example, a physics module from a science center or a robotics lab from a corporation might travel to schools and other locations.) These traveling labs will be loaded with the most modern electronic tools and, through satellite communications channels, provide learning to students in their homes, in other labs, in public or private schools, or on distance learning networks.[13] Other labs may be permanently located at sites within the community. Some may be one-room, one-person operations that serve small groups of children on an as-scheduled basis. Other labs may be larger facilities, with separate buildings or pods surrounding a central "hub." They might serve groups of students that spend weeks or even months at that lab working on larger scale learning projects in topics such as space science, marine biology, musical performance, or art. These larger learning labs may be privately operated and survive based on their quality of service and need as determined by the education marketplace.

TOMORROW'S CLASSROOM

Meet Jackie and José—students enrolled in intermediate algebra at Middletown High in 1997. On Mondays, the whole class meets with their teacher to ask questions, get a sense of where they're heading, and find out how the problems they've been solving fit into the overall curriculum.

But this is Tuesday. Jackie and José report to the "Mediated Learning Center," where students work at networked learning stations. Each one can access instruction in a variety of formats—text, hypertext, graphics, animation, simulation, visualization, and full-motion audio and video.

José begins at lesson six of the intermediate algebra curriculum, and Jackie at lesson seven. The instructional program contains a pre-test that pinpoints the best starting place for each. But they don't have to proceed in any particular order. Each student can navigate through a topic or lesson over a number of distinct pathways, at his or her own pace.

That doesn't mean it's a free-for-all. Not by a long shot. The students are working toward specific performance goals, and they can spend as much time as they need to achieve them—working on a given topic, exercise, problem, or skill-building activity. The students get feedback as they work, when it is most useful, as well as more thorough individual progress reports at regular intervals, so that they can monitor and regulate their own progress. When they need more help, they can access video clips of top-notch teachers explaining particular concepts or clearing up common misconceptions.

Jackie stays for two hours. She's always been frustrated that the standard 40-minute class ends just as she's hitting her stride. And she enjoys spending some time with the tutor who is always on-site and available for individual and small group help.

José absorbs the material better in shorter spurts, so he leaves after half an hour, with plans to return later. As he leaves the learning station, he takes along a personal academic notebook—a set of printed materials and exercises in two languages that are geared specifically to his strengths and weaknesses. It may be supplemented by videotapes or exercises on floppy disks.

Where are the teachers? They are in school, working with small groups and meeting with individual students. They are teaching courses that do not lend themselves to technology-mediated instruction. They are supervising mentors. And they are developing new curricula. Teachers remain the heart and soul of the institution, but they are freed up from many of the more mundane, repetitive instructional tasks, such as scoring tests and keeping records.

Mediated learning meets the needs of a broad spectrum of students—those who can move quickly through the materials, those who benefit from a slower pace, those for whom English is a second language, and those who have disabilities. And perhaps most importantly, it gives students more control over their learning so they can master the autonomous learning skills that are so essential in the Knowledge Age.

—***Business Week,*** *The Future of Technology in Education: Transforming the Way We Learn, Special Advertising Section (November 15, 1993).*

> ### TOMORROW'S TEXTBOOKS
>
> ... tomorrow's textbooks will be more like study guides—personal academic notebooks that accompany computer-mediated learning systems. They will provide an organizational guide, mapping out possible routes through the courseware, as well as guides for evaluating progress. They may provide homework problems and collaborative exercises for small groups to follow.
>
> For publishers, this represents a real turnabout. Instead of producing elaborate, four-color textbooks with relatively primitive electronic ancillaries, they are likely to produce mediated learning systems accompanied by guides that are flexible, easy-to-use, and less expensive to produce.
> —***Business Week,*** *The Future of Technology in Education: Transforming the Way We Learn, Special Advertising Section (November 15, 1993).*

> ### IT'S ALREADY HAPPENING
>
> Today, some pioneers, such as Jostens Learning, are offering the first generation of tomorrow's learning systems that delivers interactive, multimedia instruction over a network, using the kinds of microcomputers that many schools already own. InterActiveMedia is now being tested in schools throughout the country, and offers a glimpse of the kinds of activities that will dominate tomorrow's classroom.
> —***Business Week,*** *The Future of Technology in Education: Transforming the Way We Learn, Special Advertising Section (November 15, 1993).*

Some parents may choose to enroll their child in a particular lab where all of their educational program will be carried out at that facility. Other parents may want to provide some of their child's education at home using home computers and telecommunications equipment that will be common. Other parts of a child's education experience might be done at a smaller, more specialized learning lab, where the child will spend some time in small or large groups working on projects. Parents might be able to have a choice of particular teachers or clusters of teachers that offer a particular program that they like.

ON-LINE HOME SCHOOLING

Information technology is moving work and schooling back closer to the home. An increasing number of companies are allowing people to work at home. According to a recent lengthy analysis of American industry, the home-based workforce is growing five times as fast as other forms of employment.[14] "The breakdown of the family over the last two centuries has been driven by the rise of large corporate and government structures," observes Christopher Check, associate editor of *The Family in America,* published by the Rockford Institute. "First men were forced to leave the home to make a living, and in recent years women have followed. But information technology is moving work back closer to the home. The 'electronic cottage,' where a family can plug into networks around the globe, may restore some of the household's lost prestige as a center of economic production."[15]

The convergence of these two movements may signal a reemergence of the backwoods home school, except now the home is connected to the outside world via the Internet and is loaded to the hilt with modern (but generally affordable) electronic communications equipment. Instead of taking students to the school, the school comes to the student's home and kids receive and turn in their schoolwork by modem, tap into Congressional records, and gossip with their classmates via e-mail. Students can download classic books and artwork from Internet files, peruse university libraries, join interactive classes, and collaborate on academic projects with other children around the country and even around the world. They can study algebra, writing, and literature using well-designed instructional programs, lessons, and practice exercises. Parents have an opportunity to be involved in and observe their children's learning. Perhaps families choose to subscribe to the curriculum of the "Hartsford Community School" or the "Home-10 Learning Academy." These accredited organizations provide an array of educational materials and services, some on-line, some off-line, that are desirable to the parents, and they award diplomas that are recognized by the best colleges and universities. In addition to talking on-line with other students, students meet in groups at libraries, bookstores, museums, and parks. Some students work at apprenticeships, learning computer programming, accounting, or science at local businesses or hospitals. At least four such "cyberschools" are opening in California, Texas, and Michigan, challenging the notion that schools need walls, that home schools are for religious hermits, and that children need structured courses in order to learn.

NEW SERVICES FOR SCHOOLS AT HOME

New communication technologies make it possible for learners to study at home what previously required enrolling in a college course. America Online, for instance, currently features a home-schooling forum with lesson plan resources, tutoring, legislative updates, and networking with other home schoolers for group projects. The America Online home-school forum also features two "chat" rooms that become online classrooms with subject-area experts.
    —Joel Riemer, "Perspectives from a Home-Schooling Educator," **Educational Leadership** *(September 1994), pp. 53–54.*

Already we are seeing a significant trend toward learning at home. The National Home Education Research Institute in Salem, Oregon estimates that about 1 percent of the nation's 50 million school-age children are learning at home. That number has grown 15 percent a year since 1990. Florida, where policies generally are supportive of home schooling, has experienced a home schooling boom. While only about 16,000 of the state's 2.1 million students are home-schooled, that number has nearly tripled since 1988. Similar trends are occurring in other states. Part of what is driving this increasing popularity of home learning is its new appeal and popularity among middle class families. Once, home schooling was considered the primary domain of only a few reclusive families and rebellious hippies. But today, violence, mediocrity, a lack of values, and unforgiving bureaucracy drive many families to the point where they consider foregoing the public schools and looking at other options.

Research on school effectiveness has consistently pointed out that it is the home, after all, not the school, that has the largest effect on the educational success of the child.[16] These home influences, sometimes called "the academic curriculum of the home," refer to parent-child conversations about school and everyday events; encouragement and discussion of leisure reading; monitoring and discussion of television viewing and peer activities; and expressions of affection and interest in the child's academic and other progress as a person. So there is valid reason to emphasize the home as the child's primary learning resource.

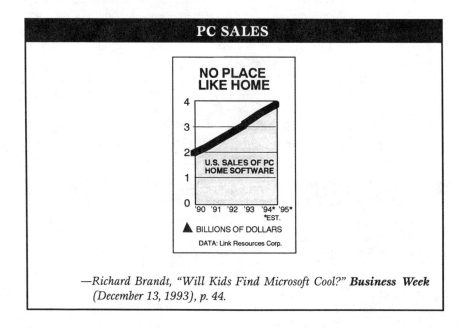

—Richard Brandt, "Will Kids Find Microsoft Cool?" **Business Week** (December 13, 1993), p. 44.

Also, the number of home-learning tools present in the home has been shown to have a substantial effect on a child's academic success. These "home learning tools" refer to things such as a typewriter, a calculator, a set of encyclopedias, and more than fifty books.[17] The availability of these mundane tools in the home apparently makes a notable difference in academic performance. So it is not difficult to imagine the potential effect that the high-powered home computer tools of tomorrow will have on children's learning. And with prices of these tools falling so rapidly, even poorer families will have access to a rich variety of learning resources.

The computer software companies are just now starting to recognize the importance of the home education market. Microsoft, the mega-size software company, recently announced that their primary target for education and communication software sales in the future will not be schools, but the home. The company has already released two major software packages specifically designed for kids: a writing program called Creative Writer and a drawing package called Fine Artist. Microsoft expects to unveil more than 100 new products within two years.[18] Other companies, such as Apple and IBM, are also moving rapidly to address the fast growing $2.8 billion home office and education market.

What is happening is that much of the best education-related software is bypassing schools and going straight into the home. In 1994, American families spent more than a half-billion dollars on education-related software, almost double the $277 million from the previous year, according to the Washington-based Software Publishers Association.[19] "It used to be that schools drove educational software. But sometime in 1994 that situation completely flip-flopped," says Lee Myers, the president of the Educational Software Institute.[20] Today, one in every three American families owns a personal computer and many of these machines are equipped with "multimedia" features, such as CD-ROM drives, vivid color animation, and stereo sound. In contrast, more than half the computers in schools are obsolescent machines such as the Apple II and early DOS models. If schools do not catch up with the home education market, they will soon come to be perceived as museums of antiquated methods of instruction.

Systems for linking the home and school are mushrooming as well. In San Jose, California, parents don't need to rely on their children to tell them what they did in school or what homework was assigned that night. They can find out by computer. Parents can also look up what the school is serving for lunch that day or check their child's attendance record. The program uses a French system known as Minitel, a five-pound machine that operates on the telephone line. The device is little more than a screen with a keyboard hookup, but it provides families with an electronic information service that has e-mail and an encyclopedia. The machines are loaned to the families for a period of time. Then, if they want to continue the service, they pay a base rate of $14.95 monthly.[21] It is likely that more innovative programs like this will emerge to link homes and schools.

Software companies are looking seriously at ideas for linking schools and homes electronically. Two California companies, Lightspan Partnership Inc. of San Diego and Computer Curriculum Corporation of Sunnyvale, are testing home-school "on-line connections." Such connections will enable teachers to send parents computer messages with daily summaries of homework assignments and allow students to collaborate on assignments from their homes.

LINKING PEOPLE TOGETHER

The futuristic thinkers and entrepreneurs of the past two decades brought us the microchip and the personal desktop computer. Now, high-tech visionaries are seeking ways to link us together as never before. Teledisc Corp, a new company founded by high-tech billionaires Bill Gates and Craig McCaw, has laid plans to link everyone, no matter where they are. Their company will create a system bringing video, voice, and data transmissions to millions of people in remote parts of the world that are bypassed by the "information superhighway." They are not the only ones with visions of a global satellite communication system. Motorola Inc., in conjunction with McDonnell-Douglas Corporation and partners in several countries, has embarked on a similar $3 billion venture called Iridium. The partnership intends to provide portable voice, fax, and paging service around the world by 1998. Hundreds of companies are providing Internet and electronic World Wide Web services.

At the same time, telephone companies are spending billions to connect people together. Pacific Telesis is building an electronic network that will connect five million homes in California. Bell Atlantic and US West are building similar electronic highways. The entire high-tech industry seems convinced that phones, TVs, and computers will one day converge into a single electronic device. People will be able to rent movies, make travel plans, take college courses, use the public library, trade stocks, watch almost any sports event, and maybe even vote—all from a living room chair. Customers using the system currently being put in place in California will have new cable boxes in their homes. They will be able to connect TVs, telephones, and computer equipment to the box and transmit information, phone calls, and video programming to and from their homes.

The nations' colleges and universities have also moved to install high-tech communication networks. About 60 percent of them already have some kind of campuswide information system or are in the process of installing one. In time, the same communications capabilities will come to pre-collegiate institutions as well.

THE INFORMATION REVOLUTION

When observers speak of the information revolution, they are referring to the widespread availability of vast volumes of data accumulated beyond the boundaries of space and time—information, both old and new, that one can acquire almost instantly from anywhere. The major developments driving this information revolution are:

- Increases in the memory capacity of computers. Terabyte memories are expected within ten years. To give you an idea of the size they are talking about here, five such memories would hold more text than the human race has ever written!
- Voice input and voice recognition systems that allow people to control machine functions via verbal commands and natural spoken speech. Although still in the early stages, developments in this area are moving rapidly. Except for typing letters or documents, the keyboard may soon be a thing of the past.
- Smaller computers. The industry's watchwords are: smaller, lighter, faster, cheaper. Nowhere is this more evident than in the emergence of handheld computers. These small handheld (or palmheld) computers have a screen you just write on with a pen. The market for pen-based computers is still tiny, with fewer than 200,000 machines in use, but it is growing rapidly, and the price declines every day. The implications of smaller and smaller electronic tools for education are profound. It is not at all unthinkable to envision students carrying small personal computers to and from schools in their backpacks or even in their shirt pockets. These computers will have capabilities that may be difficult for us to even imagine: handwriting and voice recognition, fax and e-mail transmission and reception, access to the global information networks as one walks down the street or sits in the backyard. "Throw away" computers (which are already here in the form of toys, pens, watches, and other gadgets) may also be commonplace tools for students in the future.

In addition to these developments, software designers are thinking about ways to help people find just the information they want and avoid drowning in information that is of no interest. Software gurus are now talking about *ferrets, filters,* and *mapmakers.* These are information-seeking and sorting programs that will help us avoid becoming roadkill on the information superhighway. A ferret is a program that goes out on its own exploring the world's databases for useful information. The user pre-selects which areas of information are of interest, then the ferrets go out into the various computer networks in search of the information the user is seeking. A person's computer may run hundreds of ferret programs continuously, all sep-

> ## PRODIGY LAUNCHES HOMEWORK HELPER
>
> Kids can now tap into a wealth of learning via a new on-line library database called Homework Helper. Type in the question—"What is a Kwanzaa?" and Homework Helper will search 150 newspapers, about 500 magazines, 300 works of great literature, encyclopedias, National Public Radio transcripts, and more, all updated daily, and in about five second displays a list of articles that are ranked according to the likelihood they can answer the question. Reviewers are calling the database an impressive industrial-strength research tool that parents will want as much as their kids.
>
> With built-in intelligence, Homework Helper lets you do a search by posing a question in plain English. Type in, "What is the smartest animal?" and it knows to look for articles on "clever creatures" and "intelligent beasts" as well. The system, developed by Infonautics of Wayne, Pennsylvania, is easy enough for kids to use themselves.
>
> —"Kids Can Tap into a Wealth of Learning," **USA Today** (February 9, 1995).

arately exploring the world's data for useful information. When a ferret returns, it may have to face dozens of filters that try to prevent the ferret from adding the data it found to the user's hard drive. Only data that enough filters judge to be important is passed to the mapmaker to be linked into the user's personal map of what's important, where it is, and how it relates to other information in the personal map.[22]

There are also automated-access software programs that let you pre-select the areas of a service (like CompuServe) you want to access. The program automatically logs-on to the service with your user ID and password and navigates the service for you, carrying out your assignments, finding the information you requested, and saving it as files on the computer's hard disk. Later, you can read the saved files at your leisure.

> ## PEN-BASED COMPUTERS APPEAR IN UNLIKELY WORKPLACES
>
> Teachers, nurses, and waitresses use the newfangled devices, which allow them to write on their screen. In Apopka, Florida, first-grade teacher Larrilee Rich carries the slim computer to class, tracking students' performance. Michele Hoffman, a nurse, says the machines eliminate "professional lint," the scraps of paper she would put in her pocket about her patients' progress.
>
> Many workers use the pen like a mouse, checking items off a list more often than taking notes. They find the computers make them more efficient and organized. "We do the same work now in 40% of the time we used to," says Donald Kelly, assistant state auditor in Alabama, who uses the machines to track inventories.
>
> —**Wall Street Journal**, October 18, 1994, p. 12.

Instructional Technology 205

> ### DUKE UNIVERSITY GOES PAPERLESS
>
> Using Photonics Corp's infrared network transceiver known as *Cooperative,* PowerBook computers and Farallon Computing Inc.'s *Timbuktu* communications software, Duke University implemented the paperless classroom.
>
> Engineering students are spontaneously networked with the professor and each other, making them capable of receiving instant feedback. The wireless network allows students to not only see the lectures on the big screen in front of the class, but to project their own work onto the screen.
>
> And students can continue the exchange of information outside of class using the PowerBooks and infrared transceivers.
>
> *Cooperative* enables any AppleTalk/LocalTalk-compatible devices—such as Macintosh, PowerBook, Newton and peripheral products—to communicate with one another.
>
> Approximately the size of a desktop mouse, the receiver incorporates Photonics' Infrared Transceiver technology, and data is transmitted at full LocalTalk speed through two apertures in the transceivers.
>
> Signals bounce off surfaces, making them difficult to block, according to the company.
>
> —Jeannette Brown, **Computer Reseller News** *(May 30, 1994).*

The ability to rapidly access information electronically will affect libraries as well. Most libraries already provide computer searches of their card catalogs either from within the library at terminals provided for the patrons, or from your office or home computer. A patron can also conduct computer searches of periodicals and research abstracts. In the future world of electronic information dissemination, libraries will avoid the inefficient and costly method of purchasing new books, cataloguing them, and storing them on the shelves for patrons to check out. Instead, libraries will wait until a patron requests a particular resource. Then, the library will fetch that resource over the internet. The patron may want to use the resource "on-line," or the librarian may print it out, bind it, and hand it to the patron. We may not even have to go to the library. We may be able to order the book from home by using our home computer or obtain it from a book dispenser machine (like automatic teller machines) where the user inserts a disc, has books downloaded to it, and pays for the downloaded books with a credit card.

ELECTRONIC PERFORMANCE SUPPORT TOOLS

One of the most promising areas where new high-tech tools can have a major impact on education is in the area of teacher and staff performance. Schools, like other types of organizations, have to depend on their average employees to do the job well, every time. The traditional approach to getting excellent performance from people is to provide training. This training typically takes the form of pre-school year or mid-year workshops on topics such as test construction, effective techniques for preventing and handling discipline problems, or instructional management. The impact of this type of training is minimal, because teachers are either overloaded with so much information that it is difficult for them to assimilate it all and put it into practice, or the information presented may be material that the teachers either don't need or don't need right away. When they do need it, they will probably have forgotten it. Training doesn't always result in employee competence. In most cases, training programs are stand alone "events" which leave learners essentially on their own to pursue competence. Often people forget, or there are knowledge transfer problems. It is difficult to apply the skill in the actual situation. Few people, when asked how they became competent, cite training programs or workshops as the causative factor. Most say they learned on the job by asking questions of colleagues, observing others, or through trial and error. Other traditional forms of training such as training manuals, textbooks, and conferences do not adequately provide the performance support that is required at the *moment of need*.

Businesses, especially large corporations, have identified a solution to this problem. The solution is to provide employees with computerized access to information, training, and support on demand and at the moment of need. These computer systems are called *electronic performance support systems* (EPSSs for short). They enable workers to learn what they need, when they need it. The goal of electronic performance support is to allow normally skilled employees to consistently achieve excellent results. EPSSs help people do their jobs more efficiently, productively, and independently.

Electronic performance support emerged from a business and industry community frustrated by the need to increase the competency and performance of employees and the inability of traditional training approaches to effectively and efficiently meet that need. As a result of the EPSS concept, companies like Aetna, American Express, Intel, SPRINT, Frito-Lay, and Dow Chemical are reengineering their employee training requirements around the electronic performance support concept. This concept needs to be introduced to public education.

With electronic performance support, teachers can have access to important information, techniques, help, references, and technical assistance on demand, in the format desired. Typically, the user of an EPSS can ask the system to "Remind Me" (if they just want to see a checklist or some type of memory cue to remind them about a skill); "Tell Me" (if they want to see a detailed procedure or guide); "Teach Me" (if they haven't learned this skill before, they may want to see a video demonstration, go through a formal lesson on the topic, or read an article about the skill); or the user may ask for even more detailed reference material, books, or case studies related to the problem or topic.

> ### WHO USES EPSSs?
>
> - Intel Corporation uses an EPSS to help keep the field sales force in the field (instead of always in training courses) by delivering on-the-job, up-to-the minute information and training.
> - Aetna Life & Casualty agents use an EPSS for selling Aetna's retirement products.
> - Oil companies are using an EPSS to help ensure work-site safety and protect the environment.
> - Apple Computer uses an EPSS to help developers through the process of creating print-based, instructor-led, or interactive technology based training
> - The 5000+ customer service representatives at SPRINT use an EPSS to help them provide on-line help to SPRINT customers. The system provides the service reps with instant, up-to-date access to rate and product information (including competitors' information).
>   —*Electronic Performance Support Systems, Conference Program Summaries (Dallas, TX, September, 1992).*

The Center for Educational Technology of Florida State University, as part of the Florida Schoolyear 2000 Initiative, has already developed one such EPSS for teachers of students with disabilities. The EPSS provides these teachers with instant electronic access to help, video demos, medical dictionaries, case studies, and articles about the varying disabilities of the students they deal with in their classrooms. The EPSS also helps teachers to complete the various forms that they most frequently use, write letters to parents and students, schedule events and appointments during the school day, and communicate electronically with other teachers.[23]

THE FIVE TECHNOLOGIES

Broadly, instructional technology consists of the best practices currently available for modifying human capabilities. It includes practical procedures for using computers and multimedia devices either to deliver the instruction directly to the student or to supplement portions of what is presented verbally by the teacher. Instructional technology also includes practical techniques of instructional delivery that aid effective learning, whether or not they involve the use of high-tech tools. It is a technology rich in procedures and techniques for assuring that students develop important skills and always go away eager to learn more.

Methods and tools for increasing the effectiveness of instruction and learning are known and have been available for some time. What we need to do now is apply these methods and tools in a systematic way, at the right time, in the right places, in the right way.

Clearly, developments in the power and capabilities of computers and interactive learning devices outlined in this chapter promise a different kind of learning in a more modern electronic world. Education and the social institutions we commonly call schools will take on new and different meanings as the electronic computer revolution progresses. Any educational reform, to be viable for the future, must be built on the back of this revolution. Any approach to improving learning that is not erected on a foundation of modern electronic tools is outdated and obsolete from the start.

The computer explosion and information revolution represent a major watershed for our society. As with the industrial revolution, we will experience radical changes in our societal structures and ways of doing things. The possibilities for how new electronic tools and information and communication capabilities will affect education exceed the imagination. But to be viable, these capabilities must be guided and influenced by the practices and techniques of instructional technology. Creative application of new electronic tools and capabilities does not occur by accident. It occurs by design. Without the scientific design, field-testing, and improvement process of instructional technology, new electronic tools will be nothing more than interesting gadgets—enticing and cute, but not essential to the learning process.

In addition, instructional technology, to be effective in large-scale educational reform, must be combined with the four other technologies described in this book—systems thinking, systems design, quality science, and change management. Systems thinking and systems design are the technologies that will allow us to shape our formal and informal education systems in a way that will allow them to fit productively in a new world of electronic learning resources and telecommunications. Likewise, the application of quality science to managing the activities of the programs and systems we create is critical to assuring that they meet the needs of consumers. Managers and leaders involved in moving their existing organizations toward this new future will need to be skilled in change management. Otherwise, they will not be successful in motivating and implementing such large-scale change.

Developments in computers and telecommunications will make it inevitable that parents will be able to select specific educational services and curriculum for their children. These developments will make available to the home or school a wide range of sophisticated learning tools and programs. Entrepreneurs who wish to capitalize on this trend will need to use the Five Technologies to set up systems that can go around existing education structures and reach parents and students directly. By offering customized and individualized learning programs, education innovators can reach an enormous audience that is ready and willing to pay for quality education.

FOR FURTHER READING

Instructional Technology
Gary J. Anglin (Ed.), *Instructional Technology: Past, Present, and Future,* 2nd ed. (Englewood, CO: Libraries Unlimited, 1995).
Bill Bozeman and Donna Baumbach, *Educational Technology: Best Practices from America's Schools* (Princeton, NJ: Eye on Education, 1995).
Leslie J. Briggs, Kent L. Gustafson, and Murry H. Tillman (Eds.), *Instructional Design: Principles and Applications,* 2nd ed. (Englewood Cliffs, NJ: Educational Technology Publications, 1991).
Leslie J. Briggs and Walter W. Wager, *Handbook of Procedures for the Design of Instruction,* 2nd ed. (Englewood Cliffs, NJ: Educational Technology Publications, 1981).
Patricia Ann Brock, *Educational Technology in the Classroom* (Englewood Cliffs, NJ: Educational Technology Publications, 1994).
Philip J. Brody, *Technology Planning and Management Handbook* (Englewood Cliffs, NJ: Educational Technology Publications, 1995).
Walter Dick and Lou Carey, *The Systematic Design of Instruction,* 3rd ed. (Glenview, IL: Scott, Foresman and Company, 1990).
Walter Dick and Robert A. Reiser, *Planning for Effective Instruction* (Englewood Cliffs, NJ: Prentice-Hall, 1989; 2nd ed., 1996).
Robert M. Gagné, *The Conditions of Learning,* 4th ed. (New York: Holt, Rinehart, & Winston, 1985).
Robert M. Gagné (Ed.), *Instructional Technology: Foundations* (Hillsdale, NJ: Lawrence Erlbaum, 1987).
Robert M. Gagné, Leslie J. Briggs, and Walter W. Wager, *Principles of Instructional Design,* 4th ed. (Fort Worth, TX: Harcourt Brace Jovanovich College Publishers, 1992).
Steven Hackbarth, *The Educational Technology Handbook: A Comprehensive Guide—Process and Products for Learning* (Englewood Cliffs, NJ: Educational Technology Publications, 1996).
Robert Heinich, Michael Molenda, and James D. Russell, *Instructional Media and the New Technologies of Instruction,* 4th ed. (New York: Macmillan, 1993).
Jerrold E. Kemp, Gary R. Morrison, and Steven M. Ross, *Designing Effective Instruction* (New York: Merrill, 1994).
Robert F. Mager, *Making Instruction Work* (Belmont, CA: Lake Books, 1988).
M. David Merrill with David G. Twitchell (Ed.), *Instructional Design Theory* (Englewood Cliffs, NJ: Educational Technology Publications, 1994).
Neil Postman, *Technopoly: The Surrender of Culture to Technology* (New York: Knopf, 1992).
Barbara B. Seels (Ed.), *Instructional Design Fundamentals: A Reconsideration* (Englewood Cliffs, NJ: Educational Technology Publications, 1995).
Patricia L. Smith, and Tillman J. Ragan, *Instructional Design* (New York: Merrill, 1993).

Technology Investment Discrepancy
Eric A. Hanushek, *Making Schools Work: Improving Performance and Controlling Costs* (Washington, DC: Brookings, 1994).
Henry Kelly, *Technology and the American Economic Transition* (Washington, DC: Office of Technology Assessment, 1988).

Electronic Tools and Instructional Media
Steve M. Alessi and Sidney R. Trollip, *Computer-Based Instruction: Methods and Development,* 2nd ed. (Englewood Cliffs, NJ: Prentice-Hall, 1990).
Gerald D. Bailey (Ed.), *Computer-Based Integrated Learning Systems* (Englewood Cliffs, NJ: Educational Technology Publications, 1993).
Larry Cuban, *Teachers and Machines: The Classroom Use of Technology Since 1920* (New York: Teachers College Press, 1986).
Larry Cuban, "Computers Meet Classroom; Classroom Wins," *Education Week* (November 11, 1992), pp. 36–37.
Diane M. Gayeski (Ed.), *Multimedia for Learning: Development, Application, Evaluation* (Englewood Cliffs, NJ: Educational Technology Publications, 1993).
Michael J. Hannafin and Kyle C. Peck, *The Design, Development, and Evaluation of Instructional Software* (New York: Macmillan, 1988).
Lewis J. Perelman, *School's Out: Hyperlearning, the New Technology, and the End of Education* (New York: William Morrow, 1992).
Douglas M. Towne, *Learning and Instruction in Simulation Environments* (Englewood Cliffs, NJ: Educational Technology Publications, 1995).
Royal Van Horn, "Educational Power Tools: New Instructional Delivery Systems," *Phi Delta Kappan* (March, 1991), pp. 527–533.

Electronic Performance Support Systems
Gloria J. Gery, *Electronic Performance Support Systems* (Towland, MA: Gery Performance Press, 1991).
Gloria J. Gery, *Performance Improvement Quarterly,* 8, No. 1 (1995). Special issue on electronic performance support systems.
George H. Stevens and Emily F. Stevens, *Designing Electronic Performance Support Tools: Improving Workplace Performance with Hypertext, Hypermedia, and Multimedia* (Englewood Cliffs, NJ: Educational Technology Publications, 1995).

Resources (Professional Organizations in Educational Technology)
The Association for Educational Communications and Technology (AECT). AECT is a professional umbrella organization comprised of 11 divisions that deal with different areas of interest or speciality within instructional technology. For example, the Division of Educational Media Management is concerned with the administration of media programs and media collections. The Division of Instructional Development is concerned with analysis of instructional programs and design of effective solutions. The Division of Systemic Change in Education promotes systemic change in schools and higher education through systems design approaches. AECT publishes a number of valuable resources and books dealing with aspects of instructional technology including a monthly publication, *Tech Trends,* and a research quarterly, *Educational Technology Research and Development.* Its annual convention features a major exhibition of audiovisual and computer-based hardware and software in addition to a weeklong program of educational seminars and workshops.

AECT
1025 Vermont Avenue, NW
Suite 820
Washington, DC 20005

International Society for Performance Improvement (ISPI). This organization was formerly called the National Society for Performance and Instruction (NSPI) and recently changed it name to reflect its emphasis on improving performance. ISPI is interested in the study and application of instructional technology and other performance interventions in business, industry, the military, allied health professions, government, and formal education. The society publishes a monthly journal, *Performance and Instruction,* and a research quarterly called *Performance Improvement Quarterly.*

ISPI
1300 L Street, NW, Suite 1250
Washington, DC 20005

Professional Journals in Educational Technology
Educational Technology, Educational Technology Publications, 700 Palisade Avenue, Englewood Cliffs, NJ 07632-0564.
Electronic Learning, Scholastic Inc., 730 Broadway, New York, NY 10003-9538.
Tech Trends, Association for Educational Communications and Technology, 1025 Vermont Ave. NW, Suite 820, Washington, DC 20005.
T.H.E. Journal (Technological Horizons in Education), 150 El Camino Real, Suite 112, Tustin, CA 92680-3670.
Teaching and Computers, Scholastic Inc., 730 Broadway, New York, NY 10003-9538.

NOTES

1. "U.S. Business Spends More Than $50 Billion on Training in 1994," *News & Notes* (National Society for Performance and Instruction, Washington, DC: NSPI, January, 1995), p. 6.
2. As cited in Douglas Carnine, "Testing Out 'Innovations' Might Save Money, Time, and Good Will," *Education Week* (December 8, 1993).
3. See, for example, J. A. Kulik, C. C. Kulik, and P. A. Cohen, "A Meta-analysis of Outcome Studies of Keller's Personalized System of Instruction," *American Psychologist,* 34 (1979), pp. 307-318; J. A. Kulik, C.C. Kulik, and P. A. Cohen, "Effectiveness of Computer-Based College Teaching," *Review of Educational Research,* 52 (1980), pp. 31-60; Larry Cuban, "A Fundamental Puzzle of School Reform," *Phi Delta Kappan,* 69, No. 5 (1988), pp. 341-344; Larry Cuban, "Reforming Again, Again, and Again," *Educational Researcher,* 19, No. 1 (1990), pp. 3-13; and Larry Cuban, *Teachers and Machines: The Classroom Use of Technology Since 1920* (New York: Teachers College Press, 1986).
4. Ted Sizer, "Compromises," *Educational Leadership,* pp. 34-37 (March, 1984).
5. Henry Kelly, *Technology and the American Economic Transition* (Washington, DC: Office of Technology Assessment, 1988).
6. *Ibid.*
7. An analysis of most states indicates that there is typically about one paid staff member (including administrative, instructional, and support staff) for every 10 students. Florida is a typical example. According to statistical briefs published by the Florida Department of Education, that state has approximately 200,000 paid staff for approximately 2,000,000 students.

8. Stuart Krasny, President, SK&A Research, quoted in Randy Ross, "Technology Tackles the Training Dilemma," *High Technology Business* (September, 1988). A 1991 survey conducted by *Training* magazine put the overall investment in training at $43 billion. See *Training* (October, 1991, pp. 47-49).
9. The following studies show the cost-effectiveness of computer-based instruction: Douglas Ellson, "Improving Productivity in Teaching," *Phi Delta Kappan* (October, 1986); J. D. Fletcher, David E. Hawley, and Philip K. Piele, "Costs, Effects, and Utility of Microcomputer Assisted Instruction in the Classroom," *American Educational Research Journal* (Winter, 1990); J. D. Fletcher, "Cost and Effectiveness of Computer Based Training," in *Proceedings of the 1987 IEEE Systems, Man, and Cybernetics Conference* (New York: Institute of Electrical and Electronics Engineers, Inc., 1987); J. A. Kulik *et al.*, "A Meta-Analysis of Outcome Studies of Keller's Personalized System of Instruction," *American Psychologist*, 34 (1979), pp. 307-318; J.A. Kulik *et al.*, "Effectiveness of Computer-based College Teaching," *Review of Educational Research*, 52 (1980), pp. 31-60; Cecil McDermott, *Assessment of Technology-Based Supplementary Instruction* (Cupertino, CA: Apple Computer Inc., 1987); R. P. Niemiec, M. F. Sikorski, and H. J. Walberg, "Comparing the Cost-effectiveness of Tutoring and Computer Based Instruction," *Journal of Educational Computing Research*, 5 (1989), pp. 395-408; Jesse Orlanski, "The Cost-Effectiveness of Military Training," in *Proceedings of the Symposium on the Military Value and Cost-Effectiveness of Training* (Brussels: NATO, 1985).
10. The concept of the "teacher cluster" as it relates to competition and internal co-operation is described in Charles M. Reigeluth, "The Search for Meaningful Reform: A Third-Wave Educational System," *Journal of Instructional Development*, 10, No. 4 (1987), pp. 3-14.
11. These estimates of productivity gains are from research conducted by the Morgan Stanley investment firm in New York. They are reported in William Tucker, "Industry Catches Up With the Computer Revolution, *Insight* (September 19, 1994), pp. 12-15.
12. This example is taken from William Tucker, "Industry Catches Up with the Computer Revolution, *Insight* (September 19, 1994), pp. 12-15.
13. C-SPAN already has a kind of mobile learning lab in operation—a bus loaded with video equipment, computers, and on-board information services. The bus is touring the country, visiting schools, and thereby bringing its programming resources and personnel directly to local communities' doorsteps. See "C-Span Hops on Education Bus to Deliver Resources to Schools," *Orlando Sentinel*, A-8 (November 2, 1993).
14. See William Tucker, "Industry Catches Up with the Computer Revolution," *Insight* (September 19, 1994), pp. 12-15.
15. See William Tucker, "Industry Catches Up with the Computer Revolution," *Insight* (September 19, 1994), p. 15. In addition to the growth in the home-based workforce, it also appears that an increasing number of working women are trying to find ways to stay home with their children. According to a recent article, groups like Mothers First, Mothers at Home, Home by Choice, Moms Club, the National Association of Mothers' Center, and Lawyers at Home Forum claim unprecedented growth in membership since the turn of the decade. "I sense among younger women a much more critical attitude," observes Barbara Dafoe

Whitehead, vice president of the Institute for American Values, "about all of the things that older feminists accepted as truth. There is a growing familism; we're beginning to rethink where our main values lie." Data from the Bureau of Labor Statistics that show a slower rate of increase of women in the workforce, confirm Whitehead's observation. According to bureau economist Howard Hayghe, "mothers are finding alternative methods of employment that are allowing them to stay home more with their children. They're finding part-time work, home-based businesses and freelance work." See Elena Neuman, "More Moms Are Homeward Bound," *Insight* (January 10, 1994), pp. 16–19.

16. Herbert J. Walberg, "Improving the Productivity of America's Schools," *Educational Leadership*, 41, No. 8 (May, 1984), pp. 19–30; Herbert J. Walberg, "Families as Partners in Educational Productivity," *Phi Delta Kappan*, 65, No. 6 (1984), pp. 396–400. M. E. Graue, T. Weinstein, and H. J. Walberg, "School-based Home Instruction and Learning: A Quantitative Synthesis," *Journal of Educational Research*, 76 (1983), pp. 351–360; J. L. Epstein, "Effects of Teacher Practices of Parent Involvement on Student Achievement in Reading and Math," in S. Silver (Ed.), *Literacy Through Family, Community, and School Interaction* (Greenwich, CT: JAI Press, 1987); Educational Testing Service, "The Parent Principle: Prerequisite for Educational Success," *Focus*, 26 (Princeton, NJ: ETS, 1991).
17. John E. Chubb and Terry M. Moe, *Politics, Markets, and America's Schools* (Washington, DC: Brookings Institution, 1990); viz. Table 5-8.
18. Richard Brandt, "Will Kids Find Microsoft Cool?" *Business Week* (December 13, 1993), p. 44.
19. As cited in Peter West, "Home Economics: Software Developers Look Beyond Schools to Gain Market Share," *Education Week*, 14, No. 31 (April 26, 1995), pp. 1, 10.
20. *Ibid.*
21. Michelle Quinn, "Computer Links Parents to Schools," *New York Times*, March, 30, 1994, p. B-10.
22. Apple, Microsoft, and others are experimenting with "knowbots"—software detectives that scour the networks in search of precisely and only the information that the user wants.
23. Project TREE, Center for Educational Technology, Technology Reforming Exceptional Education, Florida State University, Tallahassee, FL 32306 (brochure).

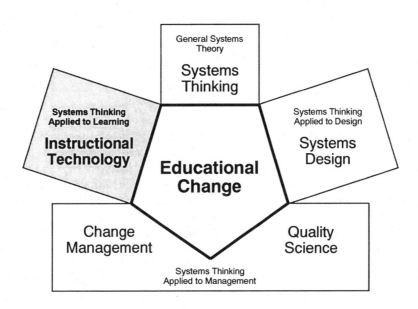

CHAPTER 11

INSTRUCTIONAL TECHNOLOGY TOOLS

Instructional technology uses scientifically based procedures to develop, evaluate, and improve instructional programs and products so that they can be optimally effective. It is concerned with the overall methodology and set of techniques employed in the instructional process. It involves detailed analysis of the subject matter to be learned, structuring the instructional presentation into an appropriate sequence, selection and use of appropriate media for presenting material, and the use of appropriate methods of assessing student performance to evaluate the effectiveness of the instruction.

Many people may still believe all that is required for good instruction is to select instructors who are "good" at their specialty. It is not necessary that these instructors know anything about communicating that specialty to others. What those who think this do not realize is that there is now a body of knowledge (called instructional technology) that is as sophisticated as any of these other specialties. This body of knowledge is about how to make instruction effective. It is a body of knowledge that is rich in procedures and techniques. In this chapter, I present some of the basic procedures and techniques that instructional technologists use in their craft.

TOOL 1: INSTRUCTIONAL DESIGN MODEL

As mentioned earlier, instructional technologists spend most of their time planning, developing, and fieldtesting instructional programs and materials. They use a process that has been developed to assure that the instructional programs and materials they create will be effective. To produce these effective programs and materials, instructional technologists follow an instructional design model. The model describes the major steps of a sound instructional design process. The insert "Educational Technology Model for Developing Effective Instruction" shows a typical instructional design model.

The diagram shown in the insert below may look a little bit complicated at first, but it is a description of the process involved in instructional technology. The first step is to determine an instructional goal—a general statement about what the learner should be able to do as a result of the instruction. The instructional goal is then expressed in a more specific way as a list of objectives. There may be several objectives for each instructional goal. After the objectives have been carefully selected and stated, tests are developed as a way to measure whether the learner is able to perform the instructional goal. The test can be used either before the instruction (to see if a learner can already perform the objectives, and if so, skip the instruction) or following the instruction (to see if the instruction was successful). Next, instructional activities are planned that will help the student learn to perform the objectives. The planning of these activities also includes selec-

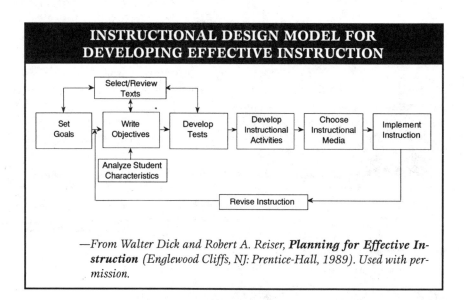

—From Walter Dick and Robert A. Reiser, **Planning for Effective Instruction** (Englewood Cliffs, NJ: Prentice-Hall, 1989). Used with permission.

tion of appropriate instructional media to be used—computer-assisted instruction, print materials, video, etc. Next, the instruction is implemented with students, usually in a field test situation. Following this, the performance of the students is examined and appropriate revisions are made so the instruction can be improved for the next group of students.

The main thing to notice about this model it that it *places responsibility for successful learning on the instruction.* In other words, if the student goes through the instruction and still cannot perform the goal, we assume that the instruction needs to be revised. This is a little bit different than the typical approach, in which we present instruction to a group of students, realizing that some will not pick up on the concepts or do well. We then discount these failures as being the fault of the students, when in reality, the instruction may have lacked clarity, may not have included appropriate examples, or may not have provided sufficient practice to allow all or even most students to acquire proficiency.

TOOL 2: PRECISION OBJECTIVES

In general conversation, we often talk about what occurs when a student learns as "understanding," "knowing," or "comprehending." But these descriptions of learning are very ambiguous and are not useful for building good instructional programs or materials. To more precisely identify the outcomes of instruction, instructional technologists write very precise statements of what the learner should be able to do upon completing a unit or module of instruction. These statements are called learning objectives. Well written learning objectives specify the materials the learner is given to work with and indicate the minimum acceptable performance by telling how well, how much, how accurate, how complete, and how fast the performance must be.

To write these objectives, instructional technologists have identified a set of action verbs that can be used to describe, with precision, the type of performance that the student should be able to do at the conclusion of the instruction. These action verbs are shown in the insert "Action Verbs for Learning Objectives." The following examples illustrate objectives with an action verb and the minimum acceptable performance.

- *Given a 3-page handwritten manuscript, the student will be able to type all of the manuscript within 20 minutes with fewer than six errors.*
- *Without the use of any materials, the student will be able to list the titles of five Shakespearean plays. All five titles must be exactly correct.*
- *Given a poorly written paragraph, the student will edit it and correct 90 percent of the grammatical errors.*

ACTION VERBS FOR LEARNING OBJECTIVES

1. Knowledge	2. Comprehension	3. Application
Recall of information	Interpret information in one's own words	Use knowledge or generalization in a new situation
arrange · name	classify · recognize	apply · operate
define · order	describe · report	choose · prepare
duplicate · recognize	discuss · restate	demonstrate · practice
label · relate	explain · review	dramatize · schedule
list · repeat	express · select	employ · sketch
match · reproduce	identify · sort	illustrate · solve
memorize	indicate · tell	interpret · use
	locate · translate	

4. Analysis	5. Synthesis	6. Evaluation
Break down knowledge into parts and show relationships among parts	Bring together parts of knowledge to form a whole and build relationships for new situations	Make judgments on basis of given criteria
analyze · differentiate	arrange · manage	appraise · evaluate
appraise · discriminate	assemble · organize	argue · judge
calculate · distinguish	collect · plan	assess · predict
categorize · examine	compose · prepare	attack · rate
compare · experiment	construct · propose	choose · score
contrast · inventory	create · set up	compare · select
criticize · question	design · synthesize	defend · support
diagram · test	formulate · write	estimate · value

—Adapted from Jerrold E. Kemp, Gary R. Morrison, and Steven M. Ross, ***Designing Effective Instruction*** (New York: Merrill, 1994), p. 85. Used with permission.

The identification and definition of precise learning objectives is an important instructional technology tool. Objectives serve as guidelines for planning the instruction and for designing the tests to be used to measure student performance. Precisely defined objectives, using appropriate action verbs, serve as a technical base from which unambiguous instruction can be derived.

Instructional Technology Tools

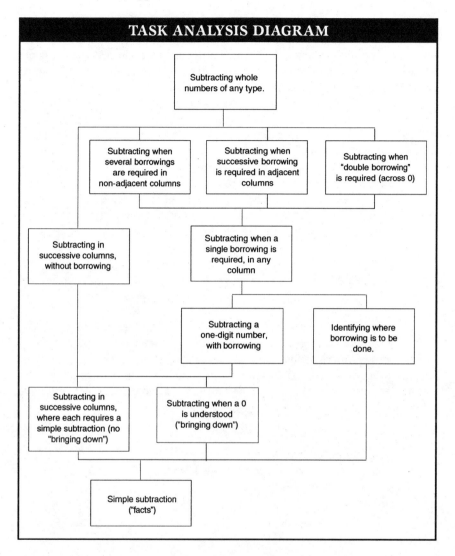

TOOL 3: TASK ANALYSIS

Another tool that instructional technologists use to develop effective instructional programs and products is *task analysis*. Task analysis is probably the most important tool the instructional technologist uses because it assures that the necessary content is included in the instruction. It also helps to assure that the content is presented in the appropriate sequence, without leaving out any necessary information. There are several different task analysis techniques that are used to define the content for an instructional module, unit, or course. These include techniques for analyzing the content structure, techniques for analyzing procedures, critical incident analysis, hierarchical task analysis, and other such techniques. Selecting the best or most appropriate technique depends on the nature of the task or content, and the environment in which the task is to be performed.

The insert "Task Analysis Diagram" shows the results of one type of task analysis. The type of analysis shown in the insert is called a hierarchical task analysis—because it shows the build-up of the components of the target skill (subtracting whole numbers). Instructional programs and products that are based on this type of scientific analysis are much more effective than instruction that is simply made up out of one's head. Even for a very experienced teacher, it is really not possible to understand how to teach a student most effectively unless one performs this type of analysis before building the instruction.

TOOL 4:
MEDIA SELECTION CHARTS

Schools rely far too heavily on the teacher, the lecture, the textbook, and supplementary print materials for presenting instructional content. Other types of instructional tools will often be better in terms of ability to provide individualized presentations and flexible amounts of practice. The following list shows other ways to deliver instruction that are very effective when used appropriately:

- self-paced learning
- computer-based tutorials
- computer-based practice and drills
- multimedia packages
- videodisc

LECTURES PLACE STUDENTS IN A PASSIVE ROLE

It is widely recognized that lectures place students in a passive role, failing to engage them in their own learning. Even students who survive such courses often absorb a very misleading impression of mathematics—as a collection of skills with no connection to critical reasoning. Yet despite its recognized ineffectiveness for most mathematics students, lecturing continues as the dominant form of instruction in mathematics classrooms.
  —National Research Council, **Moving Beyond Myths: Revitalizing Undergraduate Mathematics**, 1991.

Instructional Technology Tools

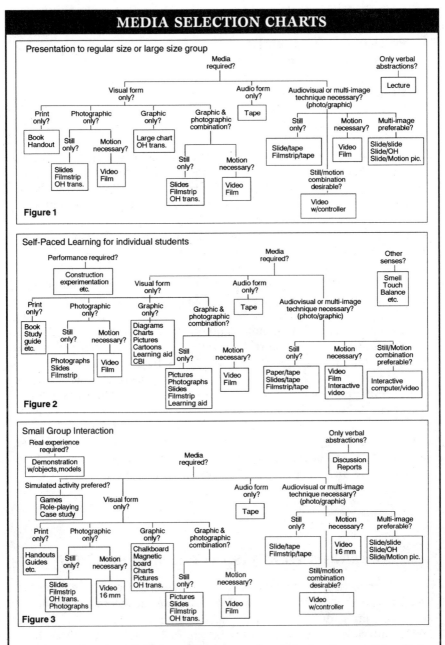

—From Jerrold E. Kemp, ***The Instructional Design Process*** *(New York: Harper & Row, 1985). Used with permission.*

- computer simulations
- audiotape with worksheets
- slidetape with worksheets
- self-instructional modules
- small-group formats
- cooperative learning
- games
- distance learning

Instructional technologists use scientific principles of learning to help determine which of the above methods should be used to deliver instruction for a particular learning objective and group of learners. They do this by analyzing what attributes need to be present in the instructional presentation. That is, they determine if pictures are needed, if self-paced practice will be needed, if display of color and motion is necessary. Then they select the most readily available and economical means that will provide the features called for by the objective.

These attributes of learning have been embodied in charts that aid the proper selection of instructional media. Examples of such charts are shown in the insert "Media Selection Charts." One chart is used for selecting media for presentation to a group. Another chart is used for selecting media for self-paced learning for individual students, and a third chart for media to be used for small group interaction. To use these charts, instructional technologists start with one or a group of related learning objectives and subject content. Then they decide on answers to the questions at the top of each chart. A "yes" answer to a question will lead to the next lower level. Then they answer the questions at that level of the chart. The decisions at each level eventually lead to a group of media from which a final choice can be made. As noted in the previous chapter, media will play a much larger role in schools and other learning environments in the future.

TOOL 5:
TESTING AND REVISING THE INSTRUCTION

Instructional technology uses scientific methodologies to test instructional programs to verify that these programs are capable of producing the desired results with real students. The goal is to identify how the instructional materials or methods might be improved. There are three basic data-gathering techniques that are used to determine the effectiveness of instruction and to locate areas for revision and improvement. The three basic techniques are pretests, embedded test items, and posttests.

Pretests

Pretesting is vitally important for trying out the instruction because it provides a benchmark to work from. Pretesting determines the learner's readiness for the instruction by showing if an individual learner already understands the subject matter or can perform some aspects of the skill to be taught. It also indicates whether the learner has appropriate preparation for starting a course or studying a topic. (For example, some students may lack basic arithmetic skills that would disqualify them from starting to study algebra.) The pretest also indicates at what point the instructional program should begin (remedial, lower level, or higher level) for that student or group of students. It also helps to motivate the learners by arousing their curiosity and interest and by showing them what will be required of them at the end of the instructional period, unit, or course. Most important, though, is that the pretest provides baseline data for determining learner growth by comparing scores on the pretest with scores on the posttest.

Pretests sometimes take the form of:

- pencil-and-paper tests;
- observations of performance (where the observer uses a rating sheet or checklist to determine competencies exhibited by the learners); or
- a questionnaire to determine background characteristics of the learners such as previous experience, interests, and goals.

For determining improvement in performance, however, the most common form of pretesting is a pencil-and-paper test based on the critical competencies to be taught. Remember, these critical competencies were expressed as *precision objectives*. Accomplishment of these critical competencies is measured by how well the learners perform on the objectives. This is why formulation of precision objectives is so important. The precision objectives form the basis for both the pretest and the posttest. The same test (or a modified form of it) can be used for both pretesting and posttesting of students.

Pretests are not used to evaluate the students or to assign grades. They are used as a way to evaluate the *effectiveness of the instructional program* that will be delivered to the students. Of interest to the instructional technologist is the amount of learning that has taken place between the pretest and the posttest. The instructional technologist wants to see if the instructional program was effective in moving the learners from the point where they cannot do something to the point where they can do it. One way to measure this is to design pretests and posttest that measure performance on precision objectives and then, after the instruction has been completed, compare pretest scores with posttest scores.

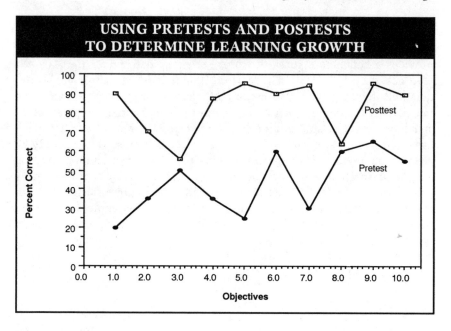

The insert "Using Pretests and Posttests to Determine Learning Growth" shows one way of laying out the pretest and posttest scores. As shown in the insert, the learning gains were large on all objectives except 3 and 8. In the case of objectives 3 and 8, only small learning gains were made. This shows that the instruction related to objectives 3 and 8 may be weak, inaccurate, or incomplete and may need to be revised and improved in some way.

Instructional technologists have various methods and techniques for constructing valid and effective pretests and posttests. These tests may be multiple-choice, short-answer, or essay type tests, or they may be tests that require observation of actual performance. The decision as to whether the pretests and posttests are multiple-choice, short-answer, or some other type is not an arbitrary decision. Rather it is based on whether the skill being tested is an intellectual skill, a conceptual skill, a verbal skill, a psychomotor skill, or, as is often the case, a combination of these. It is critical that both pretest and posttest items measure the exact competencies described in the objectives. Objectives that ask the student to *state, define, or classify* something will all require a different format for questions and responses. Sometimes teachers can construct their own tests for pretesting purposes, but often this requires the expertise of persons with in-depth training and understanding of psychometrics and other specialized knowledge.

Embedded Test Items

Embedded test items are an additional way to locate weak points in an instructional program. An embedded test item is like a practice question or problem that is part of the instructional process. The instructional technologist inserts these items into the instructional program for the purpose of seeing if the learner can perform the skill immediately after it is taught. These embedded questions may be some of the same type of questions as on the pretest and posttest but they are inserted into the instruction at appropriate points along the way. If the learner is not able to perform the skill immediately after it was taught, the instruction may need to be clarified or broken down into simpler steps or stages.

Posttests

Posttests, of course, are used to assess the objectives of the instructional module, unit, or course. The posttest may be the same test that was used as the pretest or it may be a more in-depth, more complete form of test that covers the same objectives. Posttests are generally used by teachers for the purpose of assigning a grade to each student. However, to the instructional technologist, the primary purpose of the posttest is to identify the areas of the instruction that are not working well. If a student fails to perform well on the posttest (whether the test is a pencil-and-paper test or some type of performance test), the instructional technologist should be able to look at the pretest, the embedded test items, and the posttest and identify where in the learning process the student began to fail to understand the instruction.

Once the instructional technologist has collected pretest data, data showing performance on embedded test items, and posttest data, these can be laid out in the form of a table or line graph showing how the students performed on objectives on the pretest, embedded test items, and posttest. By looking at the table or graph, the instructional technologist can see where the instruction is weak.

These are not the only tools that are used by instructional technologists, but they are samples of the kinds of tools that are used to scientifically plan, test, evaluate, and revise instruction so that it can be optimally effective for most students. Tools 6 and 7, to be discussed next, are tools that can be used by teachers, curriculum specialists, or others to locate instructional materials and programs that have been designed and tested by instructional technologists.

TOOL 6: THE EDUCATION SOFTWARE SELECTOR

The majority of instructional materials and products used by schools are "off the shelf," that is, ready-made and available from commercial companies that produce these materials for educational use. Locating the best materials for students can be a difficult task. Fortunately, there is a tool called *The Education Software Selector* (TESS). TESS is now available on CD-ROM.

TESS is produced by the Educational Products Information Exchange (EPIE) Institute. EPIE provides objective educational hardware and software evaluations to the education and training communities much like the Consumers Union provides evaluative information about household products to general consumers. EPIE's evaluations and analyses of educational materials, equipment, and systems include empirical information on performance and effects on learners. Like Consumers Union, EPIE accepts no advertising or commercial sponsorship of any kind so that it can remain unbiased. All income is derived from subscriptions, contract services from state and local education agencies, and grants. This is a comprehensive database of educational software at every level from preschool to college. TESS is the most definitive information source on availability of all types of microcomputer educational software.

EPIE also offers the following evaluation services of use to schools and school systems:

- Its newsletter *EPIEgram* covers a wide range of topics from textbooks to computer technology and is published nine times a year.
- *The EPIE Report on Computer-Based Integrated Systems* evaluates the major integrated instruction systems (or integrated learning systems). Integrated learning systems are those that include computer workstations for each student and combine hardware, software, online practice, test taking, and a management system and provide a full-blown multiyear curriculum via computer.
- EPIE's Curriculum Analysis Services for Education (CASE) provides schools with a means of analyzing, designing, and aligning its curriculum objectives, its textbooks, and other instructional resources. A school's testing program can be compared to state and national tests and curriculum priorities. The service provides grade-by-grade printed reports that are useful for understanding and improving a school's curriculum and instructional program.
- *Textbook Profiles* provides an analysis and evaluation of textbooks from an instructional technology (learning effectiveness) perspective.
- *EPIE Report* is a biannual publication that describes trends in instructional materials and equipment.
- *A-V/V PRO/FILES,* another biannual report, provides evaluative information about audiovisual and video products.
- *Micro PRO/FILES* is a bimonthly publication that gives an in depth analysis of microcomputer hardware and software.

In addition to EPIE, other agencies attempt to help educators with the computer materials selection task by conducting independent reviews and evaluations of materials. These include *MicroSIFT,* published by the Northwest Regional Education Lab, and *Pipeline,* published by the CONDUIT Clearinghouse at the University of Iowa.

TOOL 7: MULTIMEDIA INDICES

Several good indices exist to help educators select the most appropriate and effective media materials for students. One of the more comprehensive guides dealing with audiovisual media is a set of indices published by the National Information Center for Educational Media (NICEM).

The NICEM indices are arranged according to media format, such as slides, filmstrips, overhead transparencies, films, and videotapes. A CD-ROM version of these indexes is available under the title A-V On-line. This on-line version can be accessed through computerized search services such as DIALOG. Any library subscribing to such a search service will have a terminal where you can access the NICEM references on-line.

There are also other indices of instructional materials such as the one available from the National Information Center for Special Education Materials. This index provides information on the content of materials and their applicability to specific disability conditions and is helpful in preparing educational programs for handicapped learners. The Videodisc Compendium is an index devoted exclusively to videodiscs. Several other sources, such as *Core Media Collection for Elementary Schools* and *Core Media Collection for Secondary Schools,* cover a broad range of media formats These sources recommend specific audiovisual titles as core materials for elementary and secondary student learning. Other, more specialized sources or indices of media materials are listed at the end of this chapter.

FOR FURTHER READING

General References
Leslie J. Briggs and Walter W. Wager, *Handbook of Procedures for the Design of Instruction,* 2nd ed. (Englewood Cliffs, NJ: Educational Technology Publications, 1981).
Walter Dick and Lou Carey, *The Systematic Design of Instruction,* 3rd ed. (Glenview, IL: Scott, Foresman and Company, 1990).
Walter Dick and Robert A. Reiser, *Planning for Effective Instruction* (Englewood Cliffs, NJ: Prentice-Hall, 1989; 2nd ed., 1996).
Robert M. Gagné, Leslie J. Briggs, and Walter W. Wager, *Principles of Instructional Design,* 4th ed. (Fort Worth, TX: Harcourt Brace Jovanovich College Publishers, 1992).
Steven Hackbarth, *The Educational Technology Handbook: A Comprehensive Guide—Process and Products for Learning* (Englewood Cliffs, NJ: Educational Technology Publications, 1996).

Robert Heinich, Michael Molenda, and James D. Russell, *Instructional Media and the New Technologies of Instruction,* 4th ed. (New York: Macmillan, 1993).

Jerrold E. Kemp, Gary R. Morrison, and Steven M. Ross, *Designing Effective Instruction* (New York: Merrill, 1994).

Cynthia B. Leshin, Joellyn Pollock, and Charles M. Reigeluth, *Instructional Design Strategies and Tactics* (Englewood Cliffs, NJ: Educational Technology Publications, 1992).

Robert F. Mager, *Making Instruction Work* (Belmont, CA: Lake Books, 1988).

Patricia L. Smith and Tillman J. Ragan, *Instructional Design* (New York: Merrill, 1993).

Special Topics

Steven M. Alessi and Stanley R. Trollip, *Computer-Based Instruction: Methods and Development,* 2nd ed. (Englewood Cliffs, NJ: Prentice-Hall, 1991).

Benjamin S. Bloom et al., *Taxonomy of Educational Objectives: Cognitive Domain* (New York: Longman, 1956).

Larry Cuban, *Teachers and Machines: The Classroom Use of Technology Since 1920* (New York: Teachers College Press, 1986).

Michael J. Hannafin and Kyle C. Peck, *The Design, Development, and Evaluation of Instructional Software* (New York: Macmillan, 1988).

Robert F. Mager, *Preparing Instructional Objectives,* 2nd ed. (Belmont, CA: Pitman Learning, 1984).

Robert A. Reiser and Robert M. Gagné, *Selecting Media for Instruction* (Englewood Cliffs, NJ: Educational Technology Publications, 1983).

J. Michael Spector, Martha C. Polson, and Daniel J. Muraida (Eds.), *Automating Instructional Design: Concepts and Issues* (Englewood Cliffs, NJ: Educational Technology Publications, 1993).

Patty Wolfe, Michael Wetzel, Gary Harris, Tom Mazour, and Julie Riplinger, *Job Task Analysis* (Englewood Cliffs, NJ: Educational Technology Publications, 1991).

Comprehensive Media Indexes

The most comprehensive information source on audiovisual materials is the set of indexes published by the national Information Center for Educational Media (NICEM). The address is: NICEM, P.O. Box 40130, Albuquerque, NM 87196.

Core Media Collection for Elementary Schools and *Core Media Collection for Secondary Schools,* by Lucy Gregor Brown are published by R.R. Bowker. Get the latest editions.

Computer Materials Evaluations

EPIE (Educational Products Information Exchange) Institute, 103-3 West Montauk Highway, Hampton Bays, NY 11946. Tel. 516–728–9100. Fax 516–728–9229. Publishes evaluative reports on instructional materials including computer materials and hardware.

MicroSIFT, Northwest Regional Educational Lab, 300 S.W. 6th Street, Portland, OR 97204.

Pipeline, CONDUIT Clearinghouse, University of Iowa, Oakdale, Iowa City, IA 55242.

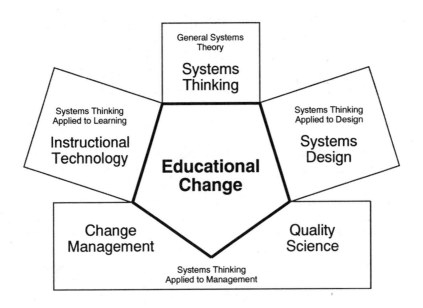

CHAPTER 12

THE FIVE TECHNOLOGIES REVISITED

In American education, the term "restructuring" often stands for nothing more than rearranging some policies at the district or school level with little resultant change in the way learning is actually pursued in the school classroom. Real change in our education systems will come about only when people (private entrepreneurs, parents, professional educators, and politicians) quit thinking about schools and schooling as what they themselves experienced and instead realize that there are numerous new and better possibilities for personal learning and academic preparation. The Five Technologies, as I have said, are the keys to unlocking these new possibilities. In addition to the Five Technologies I describe in this book, there may be additional disciplines or branches of knowledge that will be necessary to provide the proper chemistry for modernizing and refurbishing our systems of learning and education. But the Five Technologies I have written about in this book are at least the most critical ones. They each are part of the organic unity that provide the means for effective educational change.

I wrote this book because I wanted to share with you the nature of these Five Technologies. I proposed that these Five Technologies are not a mere collection of ideas and techniques but a powerful ensemble of processes, procedures, and tools that must be used together to create new and effective systems of learning and teaching. Then I described each technology separately and showed how each one contributes an essential ingredient to the total picture of educational change and reform.

In essence, these Five Technologies can be summarized as follows:

- **Systems thinking** is the technology for seeing the entire system and considering all of the factors that interact to create the results. There are certain characteristics of system behavior that can only be understood in this way. To see the entire system, we have to look at (a) internal factors such as instruction, assessment, school climate, and curriculum, and (b) external factors such as economics, market incentives, and the effect of regulations and bureaucracy. Our education systems are suffering from a lack of systems thinking. We hear frequently about proposals to redesign or reinvent schools. Yet many of these ideas are really only proposals to change one or two small aspects of the school system. They do not consider all of the internal and external factors.

- **Systems design** is a set of methods and specific activities for creating new solutions to large problems. Systems design also involves using models as a way to describe the new system. Those products and services that today make our lives comfortable and productive (such as cars, financial services, medical care, travel services, etc.) are successful because they were invented by people, companies, or entrepreneurs that understood the needs and wants of their clients and customers and used the process of systems design to meet those needs.

- **Quality science** is the technology for monitoring the processes within a system to assure that those processes produce the intended result. The people, equipment, and methods for gathering and evaluating data about the internal processes are referred to as a quality system. The quality system is missing in most educational organizations. The quality system is the subsystem that tracks data on important variables and conditions within the system. It allows students, teachers, and other employees to identify what is working and what is not. The quality system also includes the process of planning for "corrective action." Corrective action means taking actions that correct any recurring errors in the process so that the process continues to improve over time. As pointed out in this book, many of the same administrative, instructional, curriculum, and communication problems in schools occur over and over again. Corrective action means that financial and human resources are directed at fixing the root causes of routine problems to bring on positive, lasting improvements.

- **Change management** is the technology that leaders must use to guide their organizations into new realms. It allows the leader to be successful in sponsoring, initiating, and implementing change within an organization. For real change to happen, everyone in the organization must understand its significance and act in his or her role as a sponsor, advocate, agent, or target. The leader needs to use the methods noted in this volume to assess concerns and address those concerns in appropriate and positive ways. Change management doesn't tell the leader what changes ought to be made, but it does enable the leader to be successful in guiding the change through to completion.
- **Instructional technology** is part of the information and communications revolution that is driving change in almost every sector of our society today. Updating and modernizing our education systems is prerequisite to any substantial reform, restructuring, or redesign. It is only through the effective design, development, and utilization of instructional technology that we will be able to make learning effective for more and more children. The multimedia learning tools now available often are far more effective and humane for many aspects of education than are the old methods. Schools have lagged behind other enterprises in investing in modern electronic learning tools due to the perception that these electronic tools are too expensive. In actuality, they are less expensive than are the highly labor intensive and inefficient teacher presentation methods of instruction. Differentiated staffing of schools (where adults fulfill different roles in working with the children and are paid differentially according to their level of expertise and responsibility) can provide more than enough money to install the needed hardware and software. Instructional design techniques can be used to tie together all other elements into a cohesive program of instruction and learning. The incentives to make this kind of drastic change in schools have not been present due to the lack of the normal free market incentives and competition that are present in other enterprises (i.e., the private sector). Freeing the public schools from bureaucracy and monopoly will be necessary to produce real innovation and improvement in quality, productivity, and customer satisfaction.

SOME BEGINNINGS

Various education reform projects are going on around the country which, at least to some degree, are incorporating the Five Technologies. One of these is the Florida Schoolyear 2000 Initiative, a collaborative effort between The Center for Educational Technology at Florida State University and Florida's Department of Education. Schoolyear 2000 is approaching the problem of designing a new system of education by working cooperatively with nine public school districts and with a private company—Encyclopeadia Britannica Educational Corporation—as the major business partner, to create an innovative set of schools based on technology. The Florida Schoolyear 2000 Initiative is using the systems thinking and systems design approach I have described in this book as well as change management, quality science, and instructional technology. This initiative promises to be one of the major reform efforts of the next decade.

The Edison Project is another innovative approach to education systems change. Begun by Christopher Whittle, one of the nation's most ambitious businessmen and entrepreneurs, this project has set out to create a chain of innovative and modern schools pieced together from the best school ideas in the world and aimed at the average American. Edison's designers appear to be thinking about education at the systems level; they are designing a total educational experience—curriculum, assessment, administration, and so forth.

Another example of a large educational design project that is incorporating the Five Technologies is the Disney Celebration School to be located just south of Walt Disney World (Orlando, Florida) in Disney's new 5,000 acre residential community called "Celebration." The pre-K through grade 12 school located in Celebration will be owned by the local public school district but designed by Disney. Preliminary designs for the school include individual student plans, instruction based on specific blocks of skills and content rather than on "grade levels," classrooms that are designed as "learning environments" rather than traditional classrooms, use of professionals from the business and technology community to act as mentors and instructors, a balance of one-on-one, group, and individual learning and playing activities, operation of the school year round with extended school days, and use of computer-based instruction and "homelink" programs that connect each student's home with the school via computers, electronic mail, and homework advisors.

PROSPECTS FOR THE FUTURE

Legislation passed or now pending in several states is creating greater initiative and incentive for education professionals and private entrepreneurs to create new and better education systems. Fourteen states have passed legislation allowing for the creation of charter schools. These schools (which can be part of a national model or consortium, such as the Edison Project or the Coalition for Essential Schools, or just independent schools owned and operated by groups of enterprising teachers) allow for greater flexibility and freedom from the many rules and regulations that are strangling most public schools. Some charter schools are so innovative that they have spawned opposition from the traditional school monopoly and teachers' unions. One example is the Noah Webster Academy. This charter school in Michigan proposed to provide each of its students with a state-of-the-art multimedia computer and then connect these students via computer to deliver computer-based instruction via telephone line. Students, working at home, would receive lesson materials, video lectures, and assignments via computer, and could get personalized assistance from experienced teachers through toll-free telephone help lines. Over 2,000 Michigan families signed up for the school the first year. As might be expected, the Michigan Education Association and other anti-reform groups quickly vocalized their opposition and backed legal challenges in the courts.

Washington, which has generally been hostile to such innovations as choice, charter schools, and private management, may become less so. Also, Congress may be less likely to issue new regulations or otherwise interfere with state/local initiatives or changes. The funds that are currently spent on hundreds of separate, highly prescriptive, and regulated federal K–12 education programs will likely be sent home to the states in the form of block grants or tax cuts. Such a move would substitute Washington's judgment with that of states, local communities, and parents. In this way, states would be allowed to do different things. Some would move farther toward charter schools and portable scholarships or vouchers. Others would choose to fund innovative programs or research. With states experimenting with different solutions, we will be much more likely to come up with the variety and kinds of educational systems that will meet more people's wants and needs.

As we move toward these new horizons in education, we will be forced to make some long overdue decisions about the best methods for providing education to families and children. The Five Technologies are part of the essential resources that we will need for the development of new approaches for success. These new approaches can best be applied, I believe, by the private sector and by publically funded schools that are functioning in a system of competition and choice. Otherwise, there will be little incentive to make these improvements in productivity, effectiveness, and customer satisfaction.

Great opportunities lie ahead. The combination of circumstances required for major change in the education sector has presented itself. The public is discontented with the status quo in education. Thus, the knowledge and information now available about effective learning methods can come together and allow us to break from the past. The technologies described in this volume can help drive change from the regulated and bureaucratic educational systems of the past to the customer-oriented and learning-efficient systems of the future.

INDEX

A

Acceptance phase of change, 152
Ackoff, Russell, 61
Adaptation, in systems, 29–30
Advocates
    role in change, 153
    skills and abilities required by, 154
Agents
    role in change, 153
    skills and abilities required by, 154
American Association of School Administrators, Total Quality Network, 121
American Society for Quality Control (ASQC), 121
Andersen Consulting, 179
Association for Educational Communications and Technology, 210
AT&T UCS benchmarking example, 105–110
    processes usable in education, 106–108
    results, 105–106
    suggestions for schools from, 108–110

B

Backward mapping, 81–83
Banathy, Bela H., 20, 45, 57
Beckhard, Richard, 171
Beer, Stafford, 17
Benchmarking, 103–110
    AT&T UCS benchmarking example, 105–110
    done for specific reasons, 103
    partnering in, 103
    purpose of, 103
    steps to, 103–105

C

Cause and effect diagrams, 128–131
    team use recommended, 131
Cause-and-effect thinking, compared with systems thinking, 23–25

Center for Educational
    Technology, 207
Change
  See also Change management;
    Educational change
  advocate's role in, 153, 154
  agent's role in, 153, 154
  always embedded in a larger
    system, 162
  bottom-up, 165
  checklist for, 170
  clarifying, 174–174
  commitment phase of, 152
  concerns affecting individuals
    involved in, 154–156
  concerns in, 155–160
  concerns pattern in, 157
  corporate culture impacting,
    162–163
  event or process, 163–164
  impact concerns in, 156
  implementation process,
    160–161
  key roles in successful change,
    152–154
  limited capacity to assimilate,
    160
  measuring implications of, 171,
    172
  must be compatible with
    existing culture, 163
  needing careful orchestration,
    160
  preparation, acceptance, and
    commitment stages of,
    151–152
  procedures for assessing
    concerns, 157–160
  rejection of, 160
  resistance a natural response to,
    161–162
  responding to concerns,
    159–160
  role of leadership in, 164
  self concerns in, 155–156
  small and large scale, 165
  sponsorship important to, 165
  sponsors role in, 153, 154
  stages of, 151–152
  stages of concern in, 154–156
  task concerns in, 156
  successful implementation of,
    160–161
  targets role in, 154
Change commitment chart, 171,
    173–174
Change management
  See also Change; Educational
    change
  acting synergistically with other
    technologies, 9–11
  complementary and
    interdependent with other
    technologies, 9
  on the cutting edge of
    management science, 149
  dealing with resistance to
    change, 161–162
  defined, 150
  derived from systems thinking,
    5
  expecting the unexpected,
    160–161
  focus of, 10
  function of, in educational
    change, 4–5
  and limited capacity to absorb
    change, 160
  often-asked questions about,
    163–165
  origins of, 149
  outlined, 5
  patterns and forces involved in,
    148–149
  procedures for assessing
    concerns, 157–160
  summarized, 231
  as a technology, 7, 149–150

Index 237

Change management tools
  change time scale, 175–176
  checklist for change, 170
  clarifying the change, 174
  commitment chart, 171, 173–174
  levels of use scale, 176–178
  manager's resistance assessment, 171, 173
  measuring implications of change, 171, 172
  other sources for, 178–179
Change time scale, 175–176
Check, Christopher, 199
Check sheets, 134, 136
Churchman, C. West, 12
Closed systems, 25–26
Commitment phase of change, 152
Communication networks, linking people together, 202
Computer explosion, as watershed for society, 208
Computers
  See also Electronic tools
  not used effectively in schools, 189–190
Computer simulations, for systems design, 87–89
Concerns in change
  concerns pattern, 157
  impact concerns, 156
  procedures for assessing, 157–160
  responding to, 159–160
  self concerns, 155–156
  task concerns, 156
Concurrent design, 62–63
Conner, Daryl R., 151, 155
Consensus building tools, 83–84
Consensus grid, 83–84
Consensus process form, 84, 85
Control, as a function of management operations component, 70–71

Control charts, to distinguish random from special cause variation, 132–135
Core Media Collection for Elememtary Schools, 227
Core Media Collection for Secondary Schools, 227
Corporate culture, 162–163
Corrective action, assuring, 108
Cost reduction
  aided by quality science, 100–102
  electronic tools less expensive than personnel, 191–193
  quality science key to, 95
Creative Learning Exchange newsletter, 40
Cuban, Larry, 187
Customer, new role of, 110
Customer dissatisfaction, planning for corrective action, 120–121
Customer driven quality, 119
Customer satisfaction
  applying in school setting, 113–118
  information from surveys essential to improvement, 116–118
  measuring, 113–118
  quality science approach to, 115–116
  resulting from employee satisfaction, 108
  understanding customer satisfaction data, 107–108
Customer service, advantages of providing, 118–119

D

Deming, W. Edwards, 56
Design
  as a teleological activity, 46
  core ideas and values for, 58
  need for, 57

Design alternatives, 59, 66–67
Design aspirations, purposes, and requirements, 58
Design boundaries, establishing, 57–58
Design goal, diagnosing and describing, 57
Design process
  clarifying reasons for engaging in design, 57
  creating and evaluating alternative representations, 59, 66–67
  defining expectations, aspirations, purposes and requirements of system, 58
  describing future system, 59
  diagnosing and describing design goal, 57
  establishing boundaries of design, 57–58
  establishing criteria to evaluate alternatives, 59, 67
  formulating core ideas, values, and image of system, 58
  planning for development of system, 59–60
  selecting most promising alternative, 59, 67
Detail complexity, in systems, 26–27
Disney Celebration School, 232
Double-loop systems, 28
Dynamic complexity, in systems, 26–27

E

Edison Project, 3, 232
Education
  See also Schools
  to be seen as a whole system, 20
  classroom model of teaching outdated, 188–189
  costs increasing dramatically, 191, 192
  designing for, 46
  electronic performance support tools needed by, 206–207
  harm anticipated from adoption of competitive, performance-oriented model, 99–100
  industrial model of, 54–55
  lacking incentives or rewards for using technology, 193–194
  mission of, 69–70
  non-systems thinking in, 21–22
  not taking advantge of electronic tools, 186–187
  on-line home schooling, 198–202
  processes recommended from AT&T UCS benchmarking example, 106–108
  and productivity paradox, 194–196
  reducing cost of poor quality, 100–102
  seeing as a system, 65–67
  and systems thinking, 20–22
  technology investment discrepancy compared with other service industries, 189–193
Educational change
  See also Change; Change management
  absence of accountability a problem, 52
  alternatives not scientifically compared, 71
  AT&T UCS benchmarking example, 105–110
  beginnings of, 231–232

Educational change *(continued)*
  calls for, 3–4
  changes in society demanding, 49–51
  changing entire system needed, 47–48
  concurrent design used in, 63–64
  creating incentives for using technology, 194
  designing for human performance, 56–57
  fragmented mindset leading to fragmented efforts, 20–21
  idealized design used in, 61–62
  inside-view approach to, 55
  instructional technology essential to, 183–184
  insufficient user participation in, 52
  most problems caused by the system, 56–57
  need for recognition of new and better possibilities, 229
  need to change focus to learners and learning, 49
  new organizations for learning needed, 48–51
  not resulting in improved student performance, 71
  outmoded perceptions hindering new mind-set, 52
  outside-view approach needed, 55
  piecemeal efforts not successful, 11
  potential of new technologies, 49
  projects for quality improvement, 101–102
  prospects for the future, 232–233
  reactions to announced changes, 150
  seeing and responding to people as customers, 110
  small changes not sufficient, 4
  so far limited to changes in current model, 47–48
  sources for help, 121, 123
  stuck on an S^1 curve, 53–55
  suggestions for schools from AT&T UCS benchmarking example, 108–110
Educational Products Information Exchange, 185
  evaluation services avaiable from, 226
Educational reform. *See* Educational change
Einstein, Albert, 17
Electronic meeting and decision making tools, 90
Electronic performance support systems, 205–207
Electronic performance support tools, 205–207
  needed by public education, 206–207
Electronic tools
  boosting productivity while lowering costs, 191–193
  incompatible with traditional teaching, 187
  less expensive than personnel, 191–193
  not well used in schools, 186–187
Embedded test items, 224–225
Employee satisfaction, 108
EPSS, 205–207

F

Failure, celebrating and dealing with, 108
Fishbone diagrams. *See* Cause and effect diagrams

Florida Schoolyear 2000 Initiative, 62, 105, 231–232
Flowcharts, 139–142
Futures workshop, 79–81

G

Gates, Bill, 202
Goals 2000, 3
Groupware and computer design tools
  advantages of, 87
  for systems design, 84, 86–87

H

Hall, Gene, 155, 157
Hanson, Norwood Russell, 17
Harris, Reuben T., 171
Hartsford Community School, 199
Harvey, Thomas, 170
Histograms, 134, 135
Home-10 Learning Academy, 199
Home schooling, on-line, 198–202
  computer software companies responding to, 201–202
  education-related software going directly to, 201–202
  effectiveness of, 200–201
  equipment for, 199
  systems linking home and school, 201–202
  trend increasing, 200
Human performance, designing for, 56–57

I

Ideal design, as starting place for backward mapping, 81
Idealized design, 61–62
Impact concerns, 156
Industrial age, compared with information age, 50–51
Information age, compared with industrial age, 50–51
Information revolution, 203–205
  effect on libraries, 205
  ferrets, filters, and mapmakers making information available, 203–204
  technology expanding to deal with, 203–204
  as watershed for society, 208
Input/output diagrams, 74
Inputs, processes, and outputs, 72–74
  environmental constraints on, 72–73
  feedback loop needed, 74
Inspiration, 87, 88
Instructional design model, 216–217
  placing responsibility for learning on instruction, 217
Instructional technology
  acting synergistically with other technologies, 9–11
  combining with other technologies, 207–208
  complementary and interdependent with other technologies, 9
  costs decreasing, 192
  derived from systems thinking, 5
  education needing application of, 207–208
  effectiveness of programs and materials, 186
  electronic performance support tools, 205–207
  and electronic tools, 186–187
  essential to educational reform, 183–184
  focus of, 10
  function of, in educational change, 5

Index

Instructional technology *(continued)*
 incentives and rewards for adoption needed, 193–194
 and the information revolution, 203–205
 learning in the electronic world, 196–198
 linking people together, 202
 making the classroom obsolete, 188–189
 and on-line home schooling, 198–202
 outlined, 5
 productivity paradox, 194–196
 providing scientific techniques for improvement, 184–186
 summarized, 231
 as technology, 7
 and technology investment discrepancy, 189–193
 tools effective and affordable, 192–193
 used to develop instructional programs and products, 185–186
Instructional technology tools
 the Education Software Selector, 225–227
 instructional design model, 216–217
 media selection charts, 220–222
 multimedia indices, 227
 precision objectives, 217–218
 task analysis, 219–220
 testing and revising instruction, 222–225
International Society for Performance Improvement, 211

K

Key roles in change, 152–154
Knowledge base, accessed directly by students, 188

L

LearnerFirst Benchmarking 1.0, 143
LearnerFirst Process Management, 143
Learning, in the electronic world, 196–198
Leonard, George, 47
Levels of use scale, 176–178
Leverage, to make changes in systems, 27–38
Lightspan Partnership, 202

M

Malcolm Baldrige National Quality Awards, 94
 benchmarking a criterion of, 103
 for a school, 123
Management operations component of systems, 70–71
McCaw, Craig, 202
Measurement in simple terms, 107
Measures to indicate quality, 107
Media selection charts, 220–222
Mencken, H. L., 18
MicroSHIFT, 226
Minitel, 201
Mission subsystem, 68–70
 comparing stated mission with real mission, 69
 functions of, 68–70
 preventing diversion of efforts, 69–70
 selecting application of preventive programs, 70
Modeling tools for systems thinking, 35–37
Mulitmedia indices, sources of, 227
Myers, Lee, 201

N

National Center to Improve the Tools of Educators, 185
National Home Education Research Institute, 200
National Information Center for Educational Media (NICEM), 227
National Information Center for Special Education Materials, 227
National Institute of Standards and Technology (NIST), 123
New American Schools Development Corporation, 3
Newsletter for Systems Thinking in Higher Education, 39–40
Noah Webster Academy, 233

O

ODR, Inc., 151, 152, 178
Office of Technology Assessment, US Congress, 189
Olson, Jeannette R., 57
One-legged interviews, 157, 158
On-line home schooling, 198–202
Open-ended questions, 157–159
Open systems, 25–26
Organizational Development Resources Inc. (ODR), 151, 152, 178
Organizational life cycle curve, 53–55
Organizational structure, problems lying within, 56–57
Outputs, 72–74
Owens, Harrison, 53

P

Pacific Telesis, 202
Pareto, Vilfredo, 136
Pareto charts, 136–139
Pareto Principle, 136
Pipeline, 227
Placeholders, role in concurrent design, 63
Poor quality, reducing cost of, using quality science, 100–102
Posttests, 225
Precision objectives, 217–218
Preparation phase of change, 151–152
Pretests, 223–224
Price Waterhouse Change Integration Team, 178–179
Problems
  analyzing using cause and effect diagrams, 130
  approached by systems thinking, 25
  caused by systems, 56–57
  defined by relationship with other problems, 24
  Pareto charts used for categorizing and prioritizing, 137
  wrongly blamed on individuals, not faulty systems, 97–98
Processes
  all work broken down into, for quality science, 94–95
  for a system, 72–74
  checking, to improve products, 94–95
  environmental constraints on, 72–73
Productivity paradox, 194–196

Q

Quality improvement, 101–102
  flowcharting a step to, 142
Quality measures, understanding and using, 107

Index

Quality Progress, 127
Quality science
  acting synergistically with other technologies, 9–11
  adapting wording to service organizations, 99
  as an application of systems thinking, 93
  approach used in business, 93–94
  basic concepts of, 94–95
  benchmarking, 103–105
  benchmarking example, 105–110
  breaking work down into processes, 94–95
  complementary and interdependent with other technologies, 9
  customer driven quality, 119
  derived from systems thinking, 5
  designing quality control into all steps of every process, 94–95
  focus of, 10
  function of, in educational change, 4
  getting help, 121–123
  importance of, 94
  key to controlling costs, 95
  key to improving working of processes, 95
  measuring customer satisfaction, 113–118
  and new role of customer, 110
  not just a fad, 100
  objections to using in schools, 99–100
  other technologies not workable without, 95
  outlined, 5
  planning for corrective action, 120–121
  quality improvement achieved by, 100–102
  random and special cause variation in schools, 97–98
  reducing the cost of poor quality, 100–102
  summarized, 230
  as technology, 7
  variation a critical concept for, 96–97
  virtual products, 112–113
  virtual schools, 111–112
  why it pays to provide customer satisfaction, 118–119
Quality science tools, 127–142
  cause and effect diagrams, 128–131
  check sheets, 134, 136
  control charts, 132–134
  flowcharts, 139–142
  histograms, 134, 135
  Pareto charts, 136–1139
  scatter diagrams, 139

R

Random variation, 96
  control charts used to distinguish from special cause variation, 132–135
  in schools, 97–98
Recognition, as reward, 108
Reengineering, 195–196
Reigeluth, Charles M., 57
Research and development component, 71–72
Resolution not escalation, 106

S

Scatter diagrams, 139
Schools
  See also Education
  advantages of providing customer service, 118–119
  applying customer satisfaction model to, 113–118

Schools *(continued)*
  causes of drop-out problem within, 97
  customer driven quality important to, 119
  disincentives for use of technology, 194
  instructional technology making classroom obsolete, 188–189
  lagging in use of instructional technology, 184
  learning in the electronic world, 196–198
  needing to plan for corrective action on customer dissatisfaction, 120–121
  need to follow example of business and industry in restructuring, 3–4
  not adequately supplied with computers, 190–191
  not using computers effectively, 189–190
  not using electronic tools effectively, 187
  objections to using quality science in, 99–100
  problems blamed on individuals, not faulty systems, 97–98
  problems caused by the system, 56
  random and special cause variation in, 97–98
  reengineering needed, 195–196
  sharing features with businesses and industries, 99
  suggestions for, from AT&T UCS benchmarking example, 108–110
  systems linking homes with, 201–202
  tradition of teaching preventing use of electronic tools, 187
  virtual products for, 112–113
  virtual schools, 111–112
Screening of calls, 106
Self concerns, 155–156
Senge, Peter M., 9, 33
SimCity, 88
Single cause orientation, compared with systems thinking, 23–25
Single-loop systems, 28
Sizer, Ted, 189
Society, changes in, demanding educational change, 49–51
Special cause variation, 96–97
  control charts used to distinguish from random variation, 132–135
  in schools, 97–98
Special education, relying on input from individuals themselves, 108
Sponsors
  role in change, 153
  skills and abilities required by, 154
STELLA II, 37–39
Structural Thinking Experimental Learning Laboratory with Animation (STELLA), 37–39
Synergistic relationship, 10
Synergy, defined, 9
Systemic, defined, 23
Systems
  components of, 65–67
  detail complexity and dynamic complexity in, 26–27
  homeostasis in, 26
  inputs, processes, and outputs, 72–74
  management operations component controlling, 70–71
  mission subsystem, 68–70

Systems *(continued)*
  naturally occuring and created, 22–23
  open and closed, 25–26
  research and development component, 71–72
  self-regulating by adaptation or transformation, 28–30
  setting system boundaries, 26
  single-loop and double-loop, 28
  using leverage to make changes in, 27–38
Systems design
  acting synergistically with other technologies, 9–11
  advantages of electronic tools in, 87
  complementary and interdependent with other technologies, 9
  concurrent design, 62–63
  derived from systems thinking, 5
  design approaches and strategies, 60–67
  designing for human performance, 56–57
  design process, 57–60
  fear of, 51–52
  focus of, 10
  function of, in educational change, 4
  idealized design, 61–62
  inputs, processes, and outputs, 72–74
  management operations component, 70–71
  mission subsystem, 68–70
  models of systems, 63–65
  need for, 52–55
  not an intuitive process, 60
  not using existing system as starting point, 46–47
  outlined, 5
  power of, 51
  reflecting values of participants, 52
  research and development component, 71–72
  situations calling for, 47
  starting with a vision of the ideal system, 45–47
  summarized, 230
  as technology, 7
  as a teleological activity, 46
Systems design exercises, 77–79
Systems design tools, 77–89
  backward mapping, 81–83
  computer simulations, 87–89
  consensus building tools, 83–84, 85
  exercises to develop skils, 77–79
  futures workshop, 79–81
  groupware and computer design tools, 84, 86–87
  overcoming barriers to open communication and group decision making, 87
Systems diagrams, 33–35
Systems models
  complexity a problem with, 64–65
  functions of, 63–64
  new models replacing old in business and industry, 48
Systems Thinker, The, 40
Systems thinking
  acting synergistically with other technologies, 9–11
  allowing viewing of whole systems, 18
  arguing against piecemeal changes, 48
  characteristics of, 19–20
  compared with single cause orientation, 23–25

Systems thinking *(continued)*
  complementary and interdependent with other technologies, 9
  defined, 23–25
  detail complexity defined for, 26–27
  as discipline for seeing wholes, 19–20
  dynamic complexity defined for, 26–27
  and education, 20–22
  focus of, 10
  function of, in educational change, 4
  homeostasis in, 26
  leverage in, 27–28
  major concepts of, 22–30
  open and closed systems defined for, 25–26
  other technologies derived from, 5
  outlined, 5
  relating to other technologies, 18–19
  responding to problems rather than symptoms, 25
  and self-regulation in systems, 28–30
  single loop and double loop systems, 28
  summarized, 230
  system boundaries defined for, 26
  systemic defined for, 23
  systems defined for, 22–23
  as technology, 7, 18
Systems thinking tools
  modeling tools, 35–37
  other sources for, 39–40
  STELLA II, 37–39
  systems diagram, 33–35

T

Targets
  role in change, 154
  skills and abilities required by, 154
Task analysis, 219–220
Task concerns, 156
Teaching and learning, treating scientifically, 184
Technical Assistance Research Programs (TARP), 121
Technology
  change management as, 7, 149–150
  definitions of, 7–8
  instructional technology as, 7
  potential for changing education, 49
  quality science as, 7
  systems design as, 7
  systems thinking as, 7, 18
Technology investment discrepancy, 189–193
Teledisc Corp, 202
Teleological activity, design as, 46
TESS, 225–227
Testing and revising instruction, 222–225
  embedded test items, 224–225
  posttests, 225
  pretests, 223–224
The Education Software Selector (TESS), 225–227
Total Quality Management (TQM), 94
Total Quality Network, of American Association of School Administrators, 121
Transformation, in systems, 29–30

V

Variation
  cause and effect diagrams to determine sources of, 128–131
  critical concept for quality science, 96–97
  displaying in histograms, 134
  distinguishing between random and special cause, 97
  distinguishing by using control charts, 132–135
  random, 96
  random, in schools, 97–98
  special cause, 96–97
  special cause, in schools, 97–98
  understanding sources of, 96–97
Videodisc Compendium, 227
Virtual products, 112–113
Virtual schools, 111–112

W

Whittle, Christopher, 232
Whittle Communications, 3